Immobility and Medicine

Cecilia Vindrola-Padros ·
Bruno Vindrola-Padros · Kyle Lee-Crossett
Editors

Immobility and Medicine

Exploring Stillness, Waiting and the In-Between

Editors
Cecilia Vindrola-Padros
University College London
London, UK

Bruno Vindrola-Padros
University College London
London, UK

Kyle Lee-Crossett
University College London
London, UK

ISBN 978-981-15-4975-5   ISBN 978-981-15-4976-2 (eBook)
https://doi.org/10.1007/978-981-15-4976-2

© The Editor(s) (if applicable) and The Author(s), under exclusive license to Springer Nature Singapore Pte Ltd. 2021
Chapter 2 is licensed under the terms of the Creative Commons Attribution 4.0 International License (http://creativecommons.org/licenses/by/4.0/). For further details see licence information in the chapter.
This work is subject to copyright. All rights are solely and exclusively licensed by the Publisher, whether the whole or part of the material is concerned, specifically the rights of translation, reprinting, reuse of illustrations, recitation, broadcasting, reproduction on microfilms or in any other physical way, and transmission or information storage and retrieval, electronic adaptation, computer software, or by similar or dissimilar methodology now known or hereafter developed.
The use of general descriptive names, registered names, trademarks, service marks, etc. in this publication does not imply, even in the absence of a specific statement, that such names are exempt from the relevant protective laws and regulations and therefore free for general use.
The publisher, the authors and the editors are safe to assume that the advice and information in this book are believed to be true and accurate at the date of publication. Neither the publisher nor the authors or the editors give a warranty, expressed or implied, with respect to the material contained herein or for any errors or omissions that may have been made. The publisher remains neutral with regard to jurisdictional claims in published maps and institutional affiliations.

Cover illustration: © Alex Linch/shutterstock.com

This Palgrave Macmillan imprint is published by the registered company Springer Nature Singapore Pte Ltd.
The registered company address is: 152 Beach Road, #21-01/04 Gateway East, Singapore 189721, Singapore

Contents

1 Immobility and Medicine: An Introduction 1
Cecilia Vindrola-Padros, Bruno Vindrola-Padros, and Kyle Lee-Crossett

Part I Immobile Infrastructures and Enforced Waiting

2 Lists in Flux, Lives on Hold? Technologies of Waiting in Liver Transplant Medicine 15
Julia Rehsmann

3 'Being Stuck': Refugees' Experiences of Enforced Waiting in Greece 39
Pia Juul Bjertrup, Jihane Ben Farhat, Malika Bouhenia, Michaël Neuman, Philippe Mayaud, and Karl Blanchet

4 'An (Im)Patient Population': Waiting Experiences
   of Transgender Patients at Healthcare Services
   in Buenos Aires   61
   María Victoria Tiseyra, Santiago Morcillo, Julián Ortega,
   Mario Martín Pecheny, and Marine Gálvez

5 Living in 'Limbo': Immobility and Uncertainty
   in Childhood Cancer Medical Care in Argentina   85
   Eugenia Brage

Part II Embodied Stillness and Fixity

6 Embodying Immobility: Dysphoric Geographies
   of Labour Migration and Their Transformations
   in the Therapeutic Context of 'Venda' Ancestor
   Possession in Post-apartheid South Africa   113
   Vendula Rezacova

7 Liminality and the SCI Body: How Medicine
   Reproduces Stuckedness   135
   Colleen McMillan

8 Embodied Perceptions of Immobility After Stroke   155
   Hannah Stott

9 "When You Do Nothing You Die a Little Bit": On
   Stillness and Honing Responsive Existence Among
   Community-Dwelling People with Dementia   185
   Laura H. Vermeulen

10 Stories of (Im)Mobility: People Affected by Dementia
   on an Acute Medical Unit   207
   Pippa Collins

Part III Motility and (Im)Mobile Possibilities

11 Migratory Labour and the Politics of Prevention: Motility and HPV Vaccination Among Florida Farmworkers   231
Nolan Kline, Cheryl Vamos, Coralia Vázquez-Otero, Elizabeth Lockhart, Sara K. Proctor, Kristen J. Wells, and Ellen Daley

12 Living Suspended: Anticipation and Resistance in Brain Cancer   251
Henry Llewellyn and Paul Higgs

Index   273

Notes on Contributors

Jihane Ben Farhat is an epidemiologist and biostatistician at Epicentre and Médecins Sans Frontières since 2010. In recent years, Jihane has mainly worked on pediatric themes, in particular in infectious diseases (Uganda, Kenya, and Malawi), mental health (Greece and Iraq), and malnutrition (Democratic Republic of Congo and Mali).

Pia Juul Bjertrup is a Medical Anthropologist with fieldwork experience from Burkina Faso, Eswatini, and Greece. In recent years, she has worked as a Qualitative Researcher at Epicentre on a study of refugee mental health in Greece and with Médecins sans Frontières in Eswatini on HIV and Tuberculosis. Currently, Pia is a Ph.D. fellow at the Department of Anthropology, University of Copenhagen, studying care, bereavement, and death in relation to infectious and potential epidemic disease in Burkina Faso.

Professor Karl Blanchet is the Director of the Geneva Centre of Humanitarian Studies, and professor at the Faculty of Medicine, University of Geneva. Professor Blanchet has been working in health systems research at the London School of Hygiene and Tropical Medicine since

2010 and was appointed as co-director of the Health in Humanitarian Crises Centre in 2016. He has a management and public health background and brings to extensive us experience in humanitarian contexts as a relief worker and a researcher. Professor Blanchet's research focuses on resilience issues in global health, specifically in post-conflict and conflict-affected countries. He has developed innovative research approaches based on complexity science and system thinking, and is currently focused on developing and testing people-centered methodological approaches for refugee populations in the Middle East.

Malika Bouhenia is an epidemiologist with thirteen years' experience in international health. She began her career as a Program Coordinator with Médecins sans Frontières in Ivory Coast, Chad, and Yemen. Over the past seven years, she has worked in a diverse range of epidemiologist positions with Epicentre and conducted several surveys in refugee camps, South Sudan, Chad, France, and Greece. Her areas of expertise include epidemiology of infectious diseases, surveillance, and management of outbreaks.

Eugenia Brage is a postdoctoral researcher at Universidade de São Paulo, Centro de Estudos da Metrópole-Cebrap, Brazil. She has a Ph.D. in Anthropology from the Universidad de Buenos Aires, Argentina.

Dr. Pippa Collins is a Clinical Academic with an interest in older people living with physical and cognitive frailty. She received both her Masters in Clinical Research and her Ph.D. from the University of Southampton. Her background is in physiotherapy and she has a clinical and academic interest in mobility, both as a way of moving in the world and as a means of communication. She works clinically as an Advanced Clinical Practitioner in Bournemouth, UK, treating people in their own homes who have become clinically unwell or who are in social crisis.

Dr. Ellen Daley, the Associate Dean for Translational Research and Practice, is a Professor at the College of Public Health. Dr. Daley is a behavioral researcher with interests in women's health, health literacy, reproductive health, Human Papillomavirus (HPV) prevention, adolescent health, and health risk-taking behaviors.

Marine Gálvez, B.A. in Gender Studies and Political Science from Barnard College of Columbia University. Fulbright Scholar.

Paul Higgs is Professor of Sociology of Ageing at UCL where he teaches medical sociology. He has co-authored with Chris Gilleard *Cultures of Ageing: Self, Citizen, and the Body* (2000) and *Contexts of Ageing: Class, Cohort and Community* (2005). He is the co-author of *Medical Sociology and Old Age: Towards a Sociology of Health in Later Life* (2008) with Ian Jones. Professor Higgs edits the journal *Social Theory and Health* and has published widely in social gerontology and medical sociology. He is a Fellow of the UK Academy of Social Sciences and the Gerontological Society of America.

Dr. Nolan Kline is Assistant Professor of Anthropology and Co-Coordinator of the Global Health program at Rollins College. His book, Pathogenic Policing: Immigration Enforcement and Health in the US South (Rutgers), traces the multiple health-related consequences of immigration legislation and police practices in Atlanta, Georgia. His work has been funded by the National Science Foundation, and he has published several peer-reviewed articles and book chapters on im/migrant and farmworker health and human papillomavirus interventions. As an applied, medical anthropologist who uses community-based methodologies to examine the social and political determinants of health, Kline's work intersects with policy and activism.

Kyle Lee-Crossett is based in the Institute of Archaeology, UCL. He specializes in heritage studies and currently investigates institutional (im)mobilities in the context of collecting contemporary bio- and cultural diversity in public archives and museums. He is interested in how people and materials associated with diversity run into barriers in collections practice, despite diversity being seen as a mobile and inclusive force. His work explores how pathways into collections are experienced and narrated, who and what becomes stuck, and goals and aspirations for collections diversity. This book extends his work on institutional and material culture (im)mobilities to the healthcare field.

Henry Llewellyn is a medical anthropologist and research fellow in the UCL Division of Psychiatry. His work explores relationships between scientific knowledge production, medical diagnosis, treatment decision-making, clinical trials, and subjectivity. His current project examines the social and ethical implications of new medical technologies in cancer with an ethnographic focus on changing diagnostic classification and personalized medicine in brain tumors in the UK.

Dr. Elizabeth Lockhart is a postdoctoral fellow at the University of South Florida (USF). She received her Ph.D. from USF College of Public Health and her MPH from the University of Michigan. Her main research areas of interest include sexual and reproductive health prevention and care (HIV and HPV), as well as increasing access to care via community health workers and technologies. Dr. Lockhart has worked in community-engaged research for over 10 years in Michigan and in Florida.

Philippe Mayaud is a Professor of Infectious Diseases and Reproductive Health at the London School of Hygiene and Tropical Medicine. Philippe's research and teaching interests are in HIV/AIDS and sexually transmitted infections (STI) and more recently emerging infections. He has worked in many parts of Western, Eastern, and Southern Africa, but also in the Caribbean, China, and Brazil, where he has additional research projects on arbovirus epidemiology.

Colleen McMillan is Associate Professor at Renison University College, University of Waterloo. Her research interests include: the genderization of health, photovoice as a methodology in the context of illness, interprofessional education, and participatory action research with marginalized populations. Her research and teaching methods are framed by Relational Cultural Theory which emphasizes the value of connection and reciprocity.

Santiago Morcillo is Ph.D. in Social Sciences from the University of Buenos Aires (UBA). He is a researcher at the National Council on Science and Technology (CONICET), Argentina, and Professor of Social Psychology at the Sociology department (UBA). His research focuses on the intersection between sexuality, gender, masculinity, sex industry, and health, has directed the project: "Gender and sexuality in the gaze of men who pays for sex in Argentina," and co-directs "Anti-trafficking campaign in Argentina: sexual politics and discourses on prostitution."

Michaël Neuman Director of studies at Crash/Médecins sans Frontières, Michaël Neuman graduated in Contemporary History and International Relations (University Paris-I). He joined Médecins sans Frontières in 1999 and has worked both on the ground (Balkans, Sudan, Caucasus, West Africa) and in headquarters (New York, Paris as deputy director responsible for programs). He has also carried out research on issues of immigration and geopolitics. He is co-editor of *Humanitarian Negotiations Revealed, the MSF Experience* (London: Hurst and Co, 2011). He is also the co-editor of *Saving Lives and Staying Alive. Humanitarian Security in the Age of Risk Management* (London: Hurst and Co, 2016).

Julián Ortega, Ph.D in Psychology (University of Buenos Aires), Master in Labor Studies (University of Buenos Aires).

Mario Martín Pecheny is Full Professor of Political Science and Sociology of Health, University of Buenos Aires and principal researcher at the National Council on Science and Technology (CONICET), Argentina. He has extensively published on rights and politics in health, gender, and sexuality in Latin America.

Sister Sara K. Proctor is the program coordinator of the Catholic Charities, Diocese of St. Petersburg, Mobile Medical Clinic that serves migrant farmworker populations. She is a member of American Academy of Physician Assistants and has collaborated on a number of funded research studies. She has also provided trainings related to migrant health to health providers in the Tampa Bay area. In 2013, she received the Pontifical Award Pro Ecclesia et Pontifice, and in 2003, she was named Humanitarian of the year by the Florida Academy of Physician Assistants.

Julia Rehsmann is an anthropologist and postdoctoral fellow at the University of Bern and Bern University of Applied Science. As part of the research project "Intimate Uncertainties" (SNSF project 149368), she conducted research on liver transplants in Germany and examined how this field of high-tech medicine is interwoven with moral and intimate questions about "lives worth saving." In her book, she deals with the existential, technological and political dimensions of waiting in transplant medicine. Currently, she is currently working on the interdisciplinary research project "Things of Dying" (SNSF project 188869) on inpatient palliative care in Switzerland.

Vendula Rezacova is a lecturer in social anthropology at the Institute of Sociological Studies, Faculty of Social Sciences, Charles University in Prague. Toward her Ph.D. degree in sociology and social anthropology earned in 2011 from Charles University, she has conducted fieldwork in the Venda region, South Africa. Focusing on the articulation of symbolic and social transformations within traditional medicine, this research has combined the author's interests in anthropology of medicine, religion, mobility, and gender. The author is working on a book manuscript dealing with spatio-temporal relations orienting social action as these can be worked out from constituents of "Venda" personhood.

Hannah Stott is a Chartered Psychologist and Research Fellow at University of the West of England, where she was awarded her Doctorate for her thesis exploring "Embodiment, Altered Perception and Comfort After Stroke." She has a special interest in qualitative, visual, and realist research methods and is currently working on a large realist evaluation of physiotherapy services within primary care. Prior to working in academia, she has worked supporting stroke survivors and others with long-term physical and psychological conditions and is passionate about supporting individuals to improve their well-being.

María Victoria Tiseyra is a sociologist (University of Buenos Aires) and Ph.D. scholar in Sociology at the National Council on Science and Technology (CONICET), Argentina. Her interdisciplinary training combines studies into Social Politics (UBA) and a Master's degree in Women's, Gender and Citizenship studies (University of Barcelona), Spain. Her research focuses on the intersection between gender, sexuality, disability/crip/decolonial studies. Also she has done research in sexual and reproductive health.

Dr. Cheryl Vamos is an Associate Professor and a Fellow with the Chiles Center for Women, Children, and Families. She is also a Core Faculty member of the Collaborative for Research Understanding Sexual Health (CRUSH). The overall goal of Dr. Vamos' research is to facilitate the translation of maternal and child health (MCH) evidence into practice for patients, providers, and women at-large.

Coralia Vázquez-Otero is an NCI Postdoctoral Fellow in the Department of Social and Behavioral Sciences at the Harvard T. H. Chan School of Public Health and Dana-Farber Cancer Institute. She is interested in the prevention of HPV-related cancers via educating about and promoting the HPV vaccine, and the use of evidence-based and population-based interventions (policy) to prevent and improve cancer-related outcomes and eliminate health inequities. Dr. Vázquez-Otero is originally from Puerto Rico and completed her Ph.D. at the University of South Florida College of Public Health.

Laura H. Vermeulen is a senior researcher based at the University of Amsterdam, department of Anthropology. Her Ph.D. dissertation chronicles the world-making of people with dementia living on their own in the community in The Netherlands. Her research interests include phenomenology, philosophical anthropology, responsiveness, collaborative methods, politics of learning. Her current project focuses on how social networks are being done by people with dementia, their family, friends, neighbors, and care workers in long-term care. It studies how these attempts at networking care open up a space for broadening a focus on dyadic care relations toward networked forms of support in long-term care.

Bruno Vindrola-Padros is based in the Institute of Archaeology, UCL. He specializes in material culture studies and is currently exploring manifestations of (im)mobilities in the Neolithic period. His interest lies in understanding how objects have unintended consequences on human social life when they (are) break(ing). Some of these consequences relate to social re-categorizations of space, modifications of social practices through the incorporation of broken objects, and/or the development of a social care or maintenance of certain objects. He has published articles in Antiquity and has recently expanded his work into the healthcare field by publishing in BMJ Quality and Safety.

Cecilia Vindrola-Padros is a medical anthropologist working in the Department of Targeted Intervention, UCL. One of her research interests is the exploration of (im)mobilities in healthcare. She has carried out research on medical travel and has now started to explore microforms of movement involved in the delivery of care. She is the lead editor of *Healthcare in Motion: (Im)mobilities in Health Service Delivery and Access* (co-edited with Johnson and Pfister, Berghahn Books; 2018).

Dr. Kristen J. Wells is an Associate Professor of Psychology at San Diego State University. She directs the Cancer Disparities and Cancer Communication Lab and is Co-Director of the SDSU HealthLINK Center. Her graduate training was in clinical psychology and public health, followed by post-doctoral training in behavioral oncology. Her research focuses on improving the quality of healthcare delivered to underserved populations. She has led or contributed to multiple studies that developed and evaluated patient navigation interventions to improve cancer care, HIV-related care, and mental health. She also develops technological solutions to improve adherence to recommended cancer care and oral cancer medications.

List of Figures

Fig. 4.1	Urology Division	69
Fig. 4.2	Patients with appointments to be hospitalized, form a line here	70
Fig. 4.3	Waiting room	72
Fig. 8.1	Relationship between the body, motility and performed action	160
Fig. 8.2	Multifaceted experiences of immobility after stroke	176

1

Immobility and Medicine: An Introduction

Cecilia Vindrola-Padros, Bruno Vindrola-Padros, and Kyle Lee-Crossett

The social sciences have experienced a "mobilities turn" in the last two decades, which critiqued fixed and sedentary notions of social life and drew attention to the constant flows of people, ideas and objects that permeate our daily lives. This emerging field of thought proposed to study mobilities "in their own singularity, centrality and contingent determination" (D'Andrea et al. 2011, 150), creating a new, dynamic, lens through which to (re)examine social thought and practice (Soderstrom et al. 2013, vi).

Mobility forms were considered as acting in dialectical relationship with the immobile, where flows of people, information or objects might

---

C. Vindrola-Padros (✉) · B. Vindrola-Padros · K. Lee-Crossett
University College London, London, UK
e-mail: c.vindrola@ucl.ac.uk

B. Vindrola-Padros
e-mail: bruno.padros.14@ucl.ac.uk

K. Lee-Crossett
e-mail: kyle.lee-crossett.14@ucl.ac.uk

© The Author(s) 2021
C. Vindrola-Padros et al. (eds.), *Immobility and Medicine*,
https://doi.org/10.1007/978-981-15-4976-2_1

be interrupted, fixed or suspended at specific time points (Hannam et al. 2006; Urry 2007). Recent work has highlighted the importance of thinking about the relation between mobility and immobility, where movement intersects with processes that might entail episodes of transition, waiting, emptiness, uncertainty and fixity (Adey 2006; Khan 2016; Leivestad 2016; Szakolczai 2009). These moments when motion seems to be frozen in time and space (Adey 2006), when things are stuck, incomplete or in a state of transition can point to new theoretical, methodological and practical dimensions in social studies of medicine.

Social studies of health, illness and medicine have drawn from the mobilities literature to explore the flows of people, patients, medical technologies and healthcare workers through concepts such as healthscapes (Llewellyn et al. 2017), biotech pilgrimage (Song 2010), medical travel (Inhorn 2015; Sobo 2009; Holliday et al. 2019) and mobilities of wellness (Masuda et al. 2017). Considerable work has been carried out on mobile technologies in healthcare delivery, mainly in the form of mhealth or virtual care, such as telemedicine (Lupton 2018). A significant amount of attention has been placed on the study of mobility, but not always in relation to immobility (Vindrola-Padros et al. 2018). This represents a significant gap in knowledge, as episodes of standing still or "doing nothing" are critical in shaping daily life routines. As Ehn and Lofgren (2010) have argued, these mundane activities allow us to explore how habits, thoughts and feelings are culturally shaped and might even provide insight into larger and existential social issues.

In this edited volume, we bring the concept of immobility to the forefront of social studies of medicine to answer the following questions:

- How does immobility shape processes of medical care?
- How does the medical context develop forms of movement and stasis?
- What are the theoretical and methodological challenges of studying immobility in medical contexts?
- How can we change the ways in which we conceptualise and study immobility to address these challenges?

We believe that the study of immobility can make contributions to our understanding of health, illness and medicine by highlighting previously invisible processes concerning movement and inaction in health and medicine, particularly in relation to the nature of immobility, that is, how inaction can lead to the (re)constitution of people, places and things (Khan 2016). It can illustrate the unequal distribution of mobility as a resource, mainly when movement is required to maintain health (Vindrola-Padros et al. 2018), shedding light on how immobility imaginaries enable or limit certain kinds of movement and action (Salazar and Smart 2011). Finally, a study of immobility can foster methodological innovation by experimenting with new ways to capture movement and stasis, but also by using stillness or remaining in a fixed place as a method in itself (Coates 2017). All of these contributions have implications not only for our current understanding of health, illness and medicine, but also for the analysis of mobility and immobility in other areas.

"Thinking with" Immobility

The tension between mobility and immobility is made explicit in several scenarios or cases throughout the book. As we were developing the content, we encouraged contributors to think about some concepts that might help frame different manifestations of immobility such as: (1) liminality, (2) stillness, (3) emptiness, and (4) motility. The chapters are organised in relation to these different manifestations, alluding to material culture and materialities, practices, imaginaries and emotions.

Liminality has been widely used in anthropology and other disciplines to explore conditions and practices of uncertainty, in-betweenness, never-ending transition and waiting (Szakolczai 2009) and how these are (re)produced. As Horvath et al. (2015) have argued, liminality captures situations where established structures might be dislocated and hierarchies reversed, where outcomes are uncertain. These spaces of ambivalence are, nonetheless, central to sustaining social reality (Horvath et al. 2015). In medicine, the concept of liminality has been used to explore topics such as clinical staff and parents' experiences of dealing with uncertain new-born screening results (Timmermans and Buchbinder

2010), liminality and breastfeeding (Mahon-Daly and Andrews 2002), or waiting for care (Day 2015). The concept of liminality represents a useful heuristic tool to move beyond binary oppositions (as the liminal often relates to states of in-betweenness) and focus on incompleteness and processes of becoming.

The concept of stillness considers physical and imagined restrictions of movement and a sense of "stuckedness", fixity or "going nowhere" (Hage 2009). In medicine, stillness is often understood as the state of remaining motionless or stuck, which can be desired or undesired. For instance, Gask et al. (2011) have explored the relationship between depression, isolation and feeling "stuck", while the gradual loss of motion and slow succumbing into a "fixed" state has been studied in patients' experiences of disrupted motor coordination (McMillan, this volume). An interesting dimension of the concept of stillness is its potential for uncovering different mobility empowerments (or capacities for movement), where not all who desire to move will be able to do so.

Emptiness has been associated with processes where markers of certainty are eroded, leading to a sense of vacuum, where everything seems to melt into a void. According to Khan (2016), the concept of emptiness sheds light on the affective dimension of immobility, by pointing to states or emotions that might be unavailable (Deleuze and Guattari 2004). Political anthropologists have theorised emptiness in relation to empty places of power, instances of "authority vacuum, where the place of power is empty and can only be appropriated temporarily" (Wydra 2015). In medicine, feelings of emptiness have been analysed in the context of women's experiences of miscarriage (Adolfsson et al. 2004) and couple's experiences with infertility (Phipps 1993).

Motility refers to the potential for movement, aspirations for movement that are never materialised, movement imaginaries, ideas of incompleteness and never-arriving (Leivestad 2016). It allows us to explore more symbolic dimensions of immobility in the sense that movement might only rest in the imaginaries of individuals and never enacted in practice. In medicine, dimensions of motility are present in the concept of disnarration proposed by Vindrola-Padros and Brage (2017) to analyse parents' stories of their child's illness, where parents reflect on past and future scenarios that did not occur, yet are deemed important enough to be included in the story. The authors reflect on the power of these hypothetical scenarios for uncovering instances of anxiety and regret in parents' stories (Vindrola-Padros and Brage 2017).

Throughout the book, these four concepts are not developed in isolation, but interlace to formulate complex instances of waiting, restraint and the negotiation of imagined scenarios. The contributors bring these concepts to life in their rich descriptions of clinical scenarios, individuals' search for services and their struggles with ill health. The book brings together contributors from a wide range of disciplines (anthropology, sociology, social work, medicine, psychology and public health) working in countries such as Argentina, Canada, Czech Republic, Germany, Greece, South Africa, Netherlands, Switzerland, the US and the UK. These contributions highlight the importance of integrating concepts of immobility to explore experiences of enforced waiting and their effect on mental and physical health, patients' search and access to medical services for cancer treatment and transplants, and embodied experiences of spinal cord injury, stroke, cancer and dementia.

Immobile Infrastructures and Enforced Waiting

The chapters in this section of the book highlight the need to consider the invisible infrastructures of immobility, the processes, structures and actors that both enable and hinder immobile states. In doing so, however, they move beyond representations of these immobile states as passive and, instead, underscore the dynamic role of actors in maintaining and transgressing them. Bjertrup and colleagues, in their chapter on the lived experiences of refugees in Greece, argue that the act of waiting has varying degrees of activity. The authors analyse the disciplining practices of refugee camps, exploring how these create different emotions and reactions in refugees, with some choosing to leave the indefinite waiting state and others deciding to "endure the wait".

The variability of waiting experiences is also explored in the chapter by Tiseyra and colleagues, documenting the experiences of transgender patients seeking care in Argentina. According to these authors, the stigma associated with transgender in this country shapes the waiting experiences of patients who suffer continuous discrimination and humiliation. The waiting room embodies a history of exclusion and the power relations inherent to the medical gaze. Through the act of waiting, patients become disciplined into the ideal concept of the "good patient".

Processes of waiting are further unpacked in the chapter by Rehsman on patients' experiences of being on a transplant waiting list. Rehsman's analysis is granular as she considers the waiting list as "bureaucratic technology, marker of eligibility and symbol for patients' chances to receive life-saving treatment". She presents a detailed account of the material culture of waiting, the algorithm used to create the list or the telephone patients will not leave out of their sight, as this is how they will know if an organ is available for them. These tools and objects act as symbols of both mobility and immobility and are imbued with emotions such as fear and hope.

Brage centres her analysis on the concept of liminality, reflecting on the experiences of families who had to leave their place of origin to access cancer treatment. According to her, the geographic displacement as well as the uncertainty produced by the disease and treatment combine to generate a transitional state characterised by ambiguity, lack of belonging, loss of social ties, isolation and the feeling of "being out of time". The liminal states included in the narratives of the parents in her study demonstrate the central role of immobility in patient and family therapeutic itineraries.

Embodied Stillness and Fixity

This section focuses on the performative aspects of immobility, the different ways in which the body becomes still, stuck or fixed. The authors contributing to this section theorise the body in different ways, including the "object body" (Stott), the "denying body" (Řezáčová), the "undeserving body" (Kline, et al.) and the "damaged body" (Llewellyn and Higgs). These different depictions of bodily experience visualise embodiment as an active process reconfigured by the individual, which entails different stages of both movement and stasis.

When analysing the experiences of patients with spinal cord injury, McMillan asks if medicine reproduces feelings of "stuckedness". The SCI body is viewed as "problematic" in the sense that it questions the central purpose of medicine: cure a disease or fix an injury. The immobility of the injured body is a constant reminder to medicine of its failure to fulfil its main purpose, leading to negative configurations of this body that does not comply with its normative role. It signals the limitations of medical knowledge, as it fails to provide alternative ways of understanding the SCI body. In this sense, an interesting finding made by McMillan is the nuanced signs of resistance towards this dominant discourse, offering alternative conceptualisations of the SCI body where stuckedness and in-betweenness are recognised and made sense of.

The relationship between injury and disrupted mobility is further explored in the chapter by Stott on embodied perceptions of immobility after stroke. The unexpected and uncomfortable nature of the immobile body leads to, what the author has identified as, a psychological separation between body and self, which ultimately impacted on patients' identity. This separation creates intermediate bodily states that call into question dualistic notions of mind and body and, instead, represent a multi-modal definition of the body as "heavy", "unresponsive" or "untrustworthy".

The book explores immobility from the point of view of another body, one suffering from dementia. In their analysis of the lived experiences of patients with dementia, Collins and Vermeulen (in their respective chapters) argue that mobility and immobility shape the person's mental condition as well as the delivery of care. Two different states of movement interlace throughout care delivery: the immobile patient with dementia who must remain still or fixed and the mobile healthcare professional who must move to carry out their work. The movement of the person with dementia is always controlled and when this is not so, as in cases of wandering, the patient is labelled as risky or dangerous. Collins' chapter urges us to be sensitive to the different rhythms of movement, variable gradients of moving and staying still, that coincide in the same clinical environment. Vermeulen considers the benefits of focusing on instances of stillness her research participants highlighted as essential components of their lives, and visualises these as "phenomena that trigger creativity" and are capable of helping others to make sense of vulnerable moments in life.

Motility and (Im)Mobile Possibilities

This section focuses on the ways in which imagined possibilities for the future shape current processes of movement and stillness and potential strategies for accessing care. The chapters in this section generate interesting discussions on the important role of imagination in (im)mobility studies, in the form of both anticipation and alternative scenarios (Stephan and Flaherty 2019). Kline and colleagues use the concept of motility to examine the efforts of migrant farmworkers to vaccinate their children against HPV that might never materialise. While the benefits of vaccination were acknowledged by the participants in Kline et al.'s study, the structural barriers preventing their (and their children's) access to medical services meant that vaccination remained only in the imaginary, as a distant possibility.

Llewellyn and Higgs also explore the power of the imagination, but the imagined state is one of the potential immobility that might affect patients with brain tumours. The patients in their study were "living suspended", waiting for the progressive mental and physical debility generated by their disease. The authors argue that these patients go through an anticipatory loss of self, that is, "a subjectivity and an interpretation of oneself—mind and brain—produced in the intersubjective encounters between people with a brain tumour, their families, clinicians, biomedical technologies and the physical sensations that emerge as tumours develop and patients undergo intensive monitoring, surgery and therapies". In other words, patients' projected image of themselves being mentally lost and deprived of any agency in turn immobilises patients' actual daily life. This powerful interpretation of oneself represents a useful lens through which to understand the experiences of those living with disease.

Conclusion: Moving the Field of Immobility and Medicine

The motivation behind the book was to show how mobility cannot be understood under its own gaze but requires its contraposition with immobility. The perpetual pull or tension that lies between these two dimensions frames the medical treatment/healthcare practices/experiences presented in this book. While we decided to understand this tension by unpacking immobility into four different domains in the first instance (i.e. liminality, stillness, emptiness and motility), the chapter contributors have gone to great lengths to move this emerging field of thought to different, unplanned, directions.

The evident "take home" messages from these chapters relate to the need to consider the invisible infrastructures of immobility, exemplified in the waiting list that dictates movement and access to medical treatment, the refugee camp that facilitates and constrains movement. Stasis and permanence bring their own sets of challenges and can shift social practices in any direction. Immobility might seem initially framed around ideas of impotence or constraint, but being unable to move does not need to be seen as a disempowering condition. Power can be used to move others around you, and being "chosen" might mean not having to move. The "highly-mobile" might also experience barriers to care.

While there are some outstanding contributions drawing from the concept of liminality (Turner 1992), we feel there is some caution required with this elusive concept. In many of the experiences described as liminal in healthcare studies, many social relations appear to be reinforced, rather than dissolved or mitigated as the term originally suggests. Similarly, previously existing conditions of inequality often pervade and are even accentuated in so-called "transitional" contexts. The concept does seem to work rather well with highlighting certain aspects of the experiences described, such as the transformation of social roles, moments of deep reflection and critique, helping to blur the line between fixed categories and move away from binary dualisms. Nonetheless, it certainly should not be regarded as an absolute, as it runs the risk of obscuring crucial elements of social practice, as well as power relations.

The ethnographic approaches used in many of the chapters of the volume also highlight the importance of taking into consideration the interaction of gradients of mobility and immobility in the same space/bodies. This requires examining the practices of immobility at a granular level and exploring their reconfiguration in everyday life. A key aspect of this exploration of practice will be its analysis in conjunction with the affective dimensions of mobility and immobility: subjective experiences and feelings of emptiness, fear, hope and stuckedness that shape the many lives described in this book. These emotional states underscore aspects of humanity in our study of health and illness and our understanding of the diseased body and the healthcare establishment, often shaped by our experiences of waiting, exclusion, agency and everything in-between.

References

Adey, P. 2006. "If Mobility Is Everything Then It Is Nothing: Towards a Relational Politics of (Im)Mobilities". *Mobilities* 1 (1): 75–94.

Adolfsson, A., et al. 2004. "Guilt and Emptiness: Women's Experience of Miscarriage." *Health Care for Women International* 25 (4): 543–560.

Coates, J. 2017. "Idleness as Method: Hairdressers and Chinese Urban Mobility in Tokyo." In *Methodologies of Mobility: Ethnography and Experiment*, edited by A. Elliot, R. Norum, and N. Salazar, 109–128. Oxford, UK: Berghahn Books.

D'Andrea, A., et al. 2011. "Methodological Challenges and Innovations in Mobilities Research." *Mobilities* 6 (2): 149–160.

Day, S. 2015. "Waiting and the Architecture of Care." In *Living and Dying in the Contemporary World: A Compendium*, edited by Veena Das and Clara Han, 167–184. Oakland, CA: University of California Press.

Deleuze, G., and F. Guattari. 2004. *A Thousand Plateaus. Capitalism and Schizophrenia.* London: Continuum.

Ehn, B., and O. Lofgren. 2010. *The Secret World of Doing Nothing.* Berkeley, CA: University of California Press.

Gask, L., et al. 2011. "Isolation, Feeling 'Stuck' and Loss of Control: Understanding Persistence of Depression in British Pakistani Women." *Journal of Affective Disorders* 128 (1): 49–55.

Hage, G., ed. 2009. *Waiting*. Melbourne: Melbourne University Press.

Hannam, K., M. Sheller, and J. Urry. 2006. "Editorial: Mobilities, Immobilities and Moorings." *Mobilities* 1 (1): 1–22.

Holliday, R., M. Jones, and D. Bell. 2019. *Beautyscapes: Mapping Cosmetic Surgery Tourism*. Manchester: Manchester University Press.

Horvath, A., B. Thomassen, and H. Wydra. 2015. (eds.). *Breaking Boundaries: Varieties of Liminality*. UK: Berghahn Books.

Inhorn, M. 2015. *Cosmopolitan Conceptions: IVF Sojourns in Global Dubai*. Durham: Duke University Press.

Khan, N. 2016. "Immobility." In *Keywords of Mobility: Critical Engagements*, edited by Noel Salazar and Kiran Jayaram, 93–112. Oxford, UK: Berghahn Books.

Leivestad, H. H. 2016. "Motility." In *Keywords of Mobility: Critical Engagements*, edited by Noel Salazar and Kiran Jayaram, 133–151. Oxford, UK: Berghahn Books.

Llewellyn, H., et al. 2017. "Topographies of 'Care Pathways' and 'Healthscapes': Reconsidering the Multiple Journeys of People with a Brain Tumour." *Sociology of Health and Illness*. https://doi.org/10.1111/1467-9566.12630.

Lupton, D. 2018. *Digital Health: Critical and Cross-Disciplinary Perspectives*. London: Routledge.

Mahon-Daly, P., and G. J. Andrews. 2002. "Liminality and Breastfeeding: Women Negotiating Space and Two Bodies." *Health and Place* 8 (2): 61–76.

Masuda, J., et al. 2017. "Mobilities of Wellbeing in Children's Health Promotion: Confronting Urban Settings in Geographically Informed Theory and Practice." In *Children's Health and Wellbeing in Urban Environments*, edited by C. Ergler, R. Kearns and K. Witten, 207–222. London: Routledge.

Phipps, A. 1993. "A Phenomenological Study of Couples' Infertility: Gender Influence." *Holistic Nursing Practice* 7 (2): 44–56.

Salazar, N., and A. Smart. 2011. "Anthropological Takes on (Im)Mobility." *Identities: Global Studies in Culture and Power* 18: i–ix.

Sobo, E. 2009. "Medical Travel: What It Means, Why It Matters." *Medical Anthropology* 28: 326–335.

Soderstrom, O., et al. 2013. *Critical Mobilities*. London: Routledge.

Song, P. 2010. "Biotech Pilgrims and the Transnational Quest for Stem Cell Cures." *Medical Anthropology* 29 (4): 384–402.

Stephan, C., and D. Flaherty. 2019. "Introduction Experiencing Anticipation: Anthropological Perspectives." *The Cambridge Journal of Anthropology* 37 (1): 1–16.

Szakolczai, A. 2009. "Liminality and Experience: Structuring Transitory Situations and Transformative Events." *International Political Anthropology* 2: 141–172.

Timmermans, S., and M. Buchbinder. 2010. "Patients-in-Waiting: Living between Sickness and Health in the Genomics Era." *Journal of Health and Social Behavior* 51 (4): 408–423.

Turner, V. 1992 [1969]. "Liminality and Communitas." In *The Ritual Process: Structure and Anti-structure*. Chicago: Alpine Publishing.

Urry, J. 2007. *Mobilities*. Cambridge: Polity Press.

Vindrola-Padros, C., and E. Brage. 2017. "What Is Not, but Might Be: The Disnarrated in Parents' Stories of Their Child's Cancer Treatment." *Social Science & Medicine* 193: 16–22.

Vindrola-Padros, C., G. Johnson, and A. Pfister. 2018. *Healthcare in Motion: Immobilities in Health Service Delivery and Access*. New York: Berghahn Books.

Wydra, H. 2015. *Politics and the Sacred*. Cambridge: Cambridge University Press.

Part I

Immobile Infrastructures and Enforced Waiting

Part

2

Lists in Flux, Lives on Hold? Technologies of Waiting in Liver Transplant Medicine

Julia Rehsmann

Introduction: On Top of a Mountain

Getting comfortable in his armchair, Arthur took another quick glance at the mobile phone on the coffee table next to him. While I was familiar with this glance from other social contexts—just quickly checking the phone before, or even during, a conversation—this look of his carried an existential urgency and uncertain expectation that shed light on the particularities of liver transplants. 'It wasn't that long ago', Arthur sighed, 'I had this terrible nightmare; I was on top of a mountain, enjoying the fresh air and the amazing view when my phone suddenly rang'. It was his transplant clinic calling him, telling him that he had to come in immediately because they finally had a donor liver for him. 'But standing on the mountain top, it was impossible for me to make it in time to the

---

J. Rehsmann (✉)
Institute of Social Anthropology, University of Bern, Bern, Switzerland

Department of Health Professions, Bern University of Applied Science, Bern, Switzerland

hospital', Arthur said and shook his head in disbelief. It was there on this mountain top that he missed his chance for a transplant.

In this chapter, I want to use Arthur's relation to his mobile phone in his dreams and his everyday life to unpack the ambivalent relationship between mobility and immobility when waiting for a transplant. Based on ethnographic research in Germany, I focus on the time before transplantation to discuss the role technologies play in the field of liver transplants—a field that is characterized by uncertainties, urgency and delay. In this particular temporality of transplant medicine, such technologies as mobile phones become a critical feature that shapes the waiting experiences of patients like Arthur Berger[1] by mediating hope for a life-saving donor liver. Understanding waiting as a temporal experience that is shaped by hope, uncertainties and expectations, I scrutinize how the high-tech medical treatment of liver transplants works as a 'hope-generating machine' (Nuijten 2003, 16).

In what follows, I take a closer look at the crucial technological features of this hope-generating machine: how they are constituted, how they relate to each other, and how they affect patients. Inspired by Geoffrey C. Bowker and Susan Leigh Star's (2000) work on information infrastructures and classificatory practices, I discuss the waiting list as an invisible infrastructure and classifying technology of this hope-generating machine. Addressing the issue of visibility brings the materiality of lists to mind, and I delineate how waiting lists for transplants differ from common understandings of lists as a fixed, stable order, documented on paper. Furthermore, I show how these specially configured lists are based on the invisible workings of a complex computer algorithm. Handing decisions over to allocation algorithms makes it difficult to understand the workings of the waiting list and its conclusions. I contend that by being passed on to a computer program, decisions that have life and death consequences remain opaque for those affected by algorithms' calculations and assessments.

After discussing the particularities of lists in flux and the allocating algorithms they are based upon, I contend that the mobile phone becomes a critical feature in this invisible infrastructure. I argue that

---

[1] All names have been anonymized by the author.

in the context of liver transplants, the mobile phone is more than a mere communication device: it becomes an extension of the invisible and intangible waiting list. Discussing Arthur's experience of waiting for a transplant, I demonstrate how, during waiting, the mobile phone becomes a tangible manifestation of the ephemeral waiting list, a mediator of hope for a life-saving treatment, as well as a reminder of one's dependency on medical care. I show how the mobile phone transforms in this context from a mere technological communication tool used to reach people at any time and which increases mobility for those carrying it into an ambivalent marker of people's simultaneous mobility and immobility.

Methods

This chapter is informed by 13 months of in-depth ethnographic liver transplant medicine in Germany that was conducted in 2014 and 2015 and supplemented by several short field trips in 2015–2017. I explored liver transplant medicine in a variety of locations, from transplant clinics to people's home, medical conferences to patient support group meetings. This allowed me to include a wide range of perspectives, from nurses, hepatologists and transplant surgeons, through administrative personnel and journalists to patients and their relatives, while focusing particularly on patients' experiences before as well as after transplant. Ethics approval was awarded by the University of Leipzig. Informed by multiple problem-centred, narrative interviews with 29 patients and 13 medical professionals and other experts, in this chapter, I present one exemplary patient case to discuss in detail how waiting for a liver transplant is bound up with very intimate considerations concerning life and death that are contingent on such technologies as algorithms, lists and mobile phones.

Transplant Medicine as a 'Hope-Generating Machine'

With his liver failing due to cancer and cirrhosis, Arthur's fate was bound up with the medical possibilities transplant medicine in Germany had to offer. As his failing and tumorous liver would inevitably lead to his death, Arthur had placed all his hope on a donor liver—as a transplant remains the only long-term treatment available for end-stage liver failure. It is currently impossible to save everyone who needs a transplant; in Germany, every third person on the wait list for an organ dies while waiting. Part of public health care, transplant medicine in Germany is covered by the national statutory health insurance and is exclusively practised in university hospitals with specialized transplant clinics. Commissioned by the government, such national institutions as the German Medical Association (Bundesärztekammer) are crucial in defining criteria and formulating regulations regarding how organ donation and transplantation are to be practised at the national level. While the task of establishing criteria for allocating organs may seem merely a bureaucratic matter, it is these criteria that decide who in Germany should be eligible for a transplant and who should not—and thus are decisive for patients' treatment and chance of survival.

At the same time, transplant clinics and their organization, specialization and means differ significantly across the country, giving rise to a medical field shaped by particular local practices. While being a nationally regulated and locally shaped practice, German transplant medicine is also part of transnational flows of expertise and donor organs, especially as Germany is part of Eurotransplant—the non-profit organization responsible for the allocation of deceased donor organs in 8 European countries and across these countries' boundaries. With multiple organizations and institutions involved in shaping how organs in Germany are donated as well as transplanted, Arthur's desire for a liver transplant drew him into a complex web of legal, political and medical infrastructures and bureaucracies.

Bureaucracy is so ubiquitous in our lives that, as David Graeber stated, it 'has become the water in which we swim' (2015, 4). Dreaded by most, bureaucracies affect people differently, and while boredom or annoyance might ostensibly seem to be the more obvious associations with bureaucratic institutions and procedures, Monique Nuijten (2003) discusses their hope-generating characteristics. In her work on state power and its relationships with local communities in Mexico, Nuijten examines how state bureaucracies create expectations by fuelling hopes that 'everything is possible' (ibid., 16) and that 'things will be different from now on', while many of these promises remain unfulfilled. I see the features of fuelling hopes, and leaving promises unfulfilled, as constitutive elements in transplant medicine (Crowley-Matoka 2016; Kaufman and Fjord 2011; Sharp 2014) as well. With its promise to save and prolong lives, transplant medicine fuels the hopes of patients like Arthur that treatment, improvement or survival are possible. While these promises often turn out 'to be elusive' (Crowley-Matoka 2016, 150), it is the desire for more time and a better quality of life that drives people to commit to the uncertainties of this medical procedure as well as the uncertain waiting time for a transplant.

Transplant medicine in Germany relies on deceased donor organs as the main resource for liver transplants. As the death of potential organ donors is as unknown as it is unplannable and the suitability of their organs is equally uncertain, patients like Arthur wait for an indeterminate time for their hoped-for donor organ, never sure whether they will live long enough that a potential 'match' becomes available. Hope and waiting are closely related, and, as recent anthropological explorations have demonstrated, both phenomena are defined by a sense of uncertainty (Auyero 2012; Bandak and Janeja 2018; Ehn and Löfgren 2010; Elliot 2016; Hage 2009; Reed 2011). In their book on different ways of 'doing nothing' like daydreaming, routines and waiting, Billy Ehn and Orvar Löfgren (2010) point to the closely connected characteristics of hope, uncertainty, liminality and waiting. 'Above all, it is the liminality of waiting that makes it a special kind of doing nothing. In-between events can make people feel stuck, but such events can also generate new possibilities' (ibid., 67). In the context of liver transplants, the new possibility people wait for is nothing less than surviving their life-threatening disease and prolonging their lives. Nonetheless, due to the many uncertainties involved and the particular requirements people on the waiting list have to meet, patients on the wait list often feel stuck too, trapped in a limbo, not knowing how to live between pending death or potential survival.

Ghassan Hage (2009) has explored 'stuckedness' and 'immobility' in detail by emphasizing the powerful normative discourse about people's ability to endure and 'wait out' an undesired situation. Being able to 'properly' or 'best' wait out is also of critical importance to patients waiting-in-uncertainty for a donor liver—normative discourses that become most audible and visible in transplant clinics and at patient support group meetings. Hospital and bureaucratic spaces are saturated with power relations that shape people's waiting time, issues that Sophie Day (2016) and Javier Auyero (2012) have analysed in their respective work on bureaucratic waiting for healthcare or welfare benefits. Both authors draw attention to the arbitrariness and indeterminacy of people's wait when they need either medical care (Day 2016) or welfare benefits (Auyero 2012).

The 'hope-generating machine' (Nuijten 2003) of liver transplant medicine offers patients like Arthur, who would otherwise die of organ failure, a chance of survival, fuelling the hope that they may live for a couple more years or even decades. This hope unfolds in relation to an uncertain outcome in future. Receiving a transplant remains a mere possibility throughout the patients' wait, with no guarantee that the hope for treatment will be fulfilled. Because liver transplantation is a form of life-saving treatment, the uncertainty around whether it can be actualized in time prevails throughout waiting. In other words, death is always a possibility while hopefully waiting for a liver transplant. I contend that when livers fail, the resulting existential urgencies and intimate uncertainties (Strasser and Piart 2018) shape the particularities of this form of 'expectant waiting' (Elliot 2016).

In her analysis of waiting for migration in North Africa, Alice Elliot discusses 'expectant waiting' and draws attention to the 'distinctive juxtaposition of certainty and indeterminacy' (ibid., 110) in spouses' waiting to follow their migrated partners. This expectant waiting is characterized by the waiting spouses' sense of certainty about their migration to come, while it nonetheless remains unclear when this migration will be actualized, mirroring patients' experiences of waiting for a liver transplant. Migration, or in the case of my research, a donor liver, turns into a constant possibility that shapes people's everyday lives, relations and subjectivities, causing 'expectant waiting' over an indeterminate period of time. While waiting for migration and for a transplant constitutes very different forms of waiting, Elliot's notion of 'expectant waiting' draws attention to the ways in which hope and expectations shape waiting experiences and how a temporal orientation to the future affects how people live in the present.

In the context of waiting for a transplant, the outcome that people hope for is to survive their life-threatening liver failure by receiving a donor organ. A crucial step towards fulfilling this hope is getting access to the waiting list, because only those on the list are considered eligible for this high-tech and high-end medical treatment. Thus, carrying the opportunities for life-saving treatment and the hope of survival, the waiting list serves as a pivotal marker of eligibility for a liver transplant—unfolding in a space where the hopeful promises of transplant medicine meet healthcare infrastructures, legal frameworks and questions of triage.

Lists in Flux

Due to its centrality in allocating potentially life-saving donor livers, the waiting list requires closer consideration when thinking about technologies of waiting. Waiting lists are a central feature in German transplant medicine. They are omnipresent in patients' and physicians' narrations; they are central to medico-legal frameworks; they are mentioned in newspaper articles about the decreasing numbers of organ donations; and they played a critical role in the German transplant scandal.[2] There are multiple lists and lists that overlap. Each of the 50 transplant centres in Germany manages its own waiting list, while these individual lists are also combined in an international data pool administered by

---

[2]The so-called German transplant scandal refers to physicians' wrongful tampering with patients' medical files to increase their chances of a transplant (Connolly 2013; Shaw 2013). By doing so, physicians manipulated the waiting list and interfered with patients' chance of survival. A wide-ranging matter across several clinics and transplant programmes, the scandal has led to changes in the legal regulations and structural requirements of transplant programmes in Germany.

Eurotransplant in Leiden, the Netherlands. Only people registered in Eurotransplant's data pool are considered eligible for transplantation, and it is only among them that the scarce but vital resource of donor livers is allocated.

The allocation of donor livers does not operate on a 'first come, first served' principle but is based on a complex algorithm that generates 'match lists' for each donated organ that becomes available. To get into this data pool, or to 'get listed', is a crucial step towards receiving a transplant—without it, receiving a transplant in a German clinic is legally impossible. Although the notion of 'getting on the list' concerns first and foremost the inclusion of patients' data in Eurotransplant's data pool, no patient or doctor I encountered ever talked about data pools and algorithms. Patients and medical professionals talked about the waiting list, mostly how to get on the list and how to stay on it. Exactly how this list worked was of no interest to those facing a life-threatening liver disease; what they cared about was getting *on* the list. 'Getting on the list' was communicated by medics as the defining step on the patients' way to a transplant. While it did not mean that patients *would* get a transplant, getting on the list gave them a *chance* of a transplant.

Thus, the moment Arthur Berger had passed all necessary medical examinations and tests for a liver transplant and was considered eligible for this form of treatment, his data was forwarded to Eurotransplant and added to the international data pool administered by the non-profit organization. From that day onwards, whenever a donor liver in the Eurotransplant region became available, his data was assessed alongside that of numerous other patients' in that region to find the 'perfect match' for the available donor liver. At the same time, Arthur's clinic administered their own waiting list, of which he was part. The main reason for such local lists is that clinics are sometimes 'offered' a donor organ directly and have the opportunity to choose a suitable 'match' from among those on their own wait list. This is the case when an organ is either repeatedly declined by patients, transplant surgeons consider it unsuitable for their patients, or the organ does not meet the highest criteria and is designated as a so-called marginal organ. Patients like Arthur, whose chance of a match is limited by their age and progressive cancer are often asked whether they would also be willing to accept a 'marginal organ' as it significantly increases their chance for a transplant.

To make the process even more complex, the allocation of donor livers is not only based on medical criteria, but also based on ethical principles (who has the best chance of survival vs. who needs a liver most urgently), patients' body dimensions (a liver of a certain size will not fit in just any body or abdomen, but donor and recipient have to be of similar size) as well as spatial and temporal distances between clinics (to allow a quick transfer of the donor liver to its recipient). To keep the organ as healthy as possible for transplantation, this process takes place under immense time pressure, necessitating quick decisions and immediate responses. Furthermore, because of the inevitable multiple back-and-forth telephone calls between countries, clinics, medical professionals and waiting patients, those on the waiting list for a transplant have to be reachable day and night 24/7. Thus, telephones, and particularly mobile phones, become critical technologies for the unfolding of transplant trajectories, particularly so for patients on the waiting list.

Unlike the digital infrastructures that these lists are now part of, lists per se are nothing new: they are among the oldest written human documents (Goody 1977, 74–111). In his investigation of literacy, Jack Goody considers lists predominantly as a tool of cataloguing, encompassing such household inventories as cattle, as well as lists of deities, kin and servants. In a general sense, generating a list creates order and reduces complexities, and sometimes the created order is hierarchical, prioritizing some things over others. Lists are a 'mode of classifying' (Goody 1977, 103) and thus intrinsically linked to processes of inclusion and exclusion. Lists make decisions visible about whom or what should or should not be put on the list. The reasoning behind what gets on a list and what does not is not always entirely clear; nevertheless, lists strive to produce the illusion that everything can be easily classified. As waiting lists for transplants are also the result of classifying practices, giving some the chance of treatment and excluding others based on medico-legal regulations, they furthermore have to be understood as a triaging tool.

These cataloguing lists that Goody is writing about are defined by their materiality and visibility and the boundaries that come with their material features. They have clear beginnings and ends; they can be read 'both sideways and downwards, up and down, as well as left and right' (Goody 1977, 81). Waiting lists for transplants lack this visibility and materiality, although they present these features as snapshots in the form of temporary match lists. But these lists remain fleeting and in constant flux. Moreover, visibility is difficult to define in the context of transplant medicine with its short-lived match lists and digital data pools. Patients listed for a transplant are indeed visible to the allocating system as well as to the people administering these data sets and temporary lists. But it is only the bureaucratic, administrative aspect of this high-tech medical treatment that has an overview of all those 'on the list' and waiting for an organ. What becomes visible to the public are a wide range of numbers as well as medical conditions published in the annual reports of Eurotransplant, the German Association for Organ Donation (DSO) and local clinics. However, those who are actually waiting do not know who is waiting along with them.[3]

Waiting lists for a liver transplant are not worked through from the top-down, by crossing off one name after another. A clearly defined list with an obvious order does not exist in this field. What do exist are digital data sets and organ-specific algorithms of allocation that create temporary waiting lists (Amelang 2014, 21). While there is a pool of patients waiting for this form of treatment, concrete 'match lists' are formatted only if an organ becomes available and just for that one organ. The next donor liver most probably prompts the generation of a very different list. Patients' imaginings of 'getting on' the waiting list differ from the reality of accessing liver transplants by means of being added to an international data pool. Patients as well as physicians describe this stage as 'getting on the list', despite there being no actual list—at least in Goody's (1977) terms. Thus, the issue of clear readability and ordering that Goody (1977) discusses appears to be far from the waiting lists in liver transplant medicine, whose formation remains opaque for those whose name might appear under very particular circumstances on such a list.

---

[3]This is the case unless patients on the waiting list get in touch with patient associations or local support groups, which offer a platform for information and emotional support. Nonetheless, there tend to be very few patients who are waiting for a transplant at these meetings, as attendees are usually people who already received a transplant.

In a strict sense, waiting lists are the result of an algorithm cross-referencing a particular donor organ with all the relevant information available about patients in Eurotransplant's data pool. When a donor organ becomes available, algorithms generate a match list and the organ is placed with a recipient in accordance with that organ-specific match list of potential organ recipients. Thus, while patients tend to think that the transplant system works to find an organ *for them*, it is rather that the allocating system looks for the best possible match for each organ that becomes available. Forming anew with every organ that becomes available, these 'match lists' do not hold their shape for long. Unlike a clearly defined document listing names, they are fleeting and in constant flux.

Allocating Algorithms

Due to its centrality in creating 'match lists' and thus deciding the fate of patients like Arthur, I want to take a closer look at the algorithm that is so decisive in the allocation of donor livers as well as the numbers and scores it generates. With the aim of making the allocation as fair and impartial as possible, the decision of which donor liver should be offered to whom is delegated to an algorithm that cross-references such varied data as blood type, age, height and body weight of donor and recipient to a make a match. 'Making a match is ultimately no more than pressing a button', the voice-over explains in Eurotransplant's (2015) company film, with 'advanced computer systems' behind that 'button'. After all, who or what else could make the allocation of livers—and triaging of lives—fairer and more impartial then algorithms calculating numbers?

Highlighting the moral and political features of classifications, Bowker and Star argue that classifying schemes are 'ordinarily invisible' (2000, 2) and so ubiquitous in our everyday lives that they tend to get overlooked and taken for granted. Moreover, they point out how an increasing amount of data is organized in such highly technological and complex ways that it has become ever more challenging to trace how classifications and categories are defined, configured and employed. This is relevant to the system of transplant medicine as well, as it is a field that merges medical, political and technological terrains (Kierans and Cooper 2011).

This invisibility becomes especially problematic when the complexities of 'decision making are hidden away inside a piece of technology or in a complex representation' (Bowker and Star 2000, 135)—like allocation algorithms.

As medical anthropologist Katrin Amelang (2014, 23–24) has pointed out, algorithms in this context represent a political compromise rather than a purely objective tool for allocation. Decisions about what kind of factors should be considered relevant in this algorithm are the product of negotiations and discussion. This is particularly so in the context of a non-profit organization that allocates organs nationally and internationally and whose member states contribute to the pool of available donor organs very differently in terms of numbers. Germany, for example, receives more donor organs from the shared pool than it contributes—in contrast to other Eurotransplant member states.[4]

Numbers are decisive in this field, and they are 'trusted' (Porter 1996); they decide whether a patient is too old or sick to be considered a potential transplant recipient or organ donor. They determine whether a liver is too fat, too sick, tumours too big or too numerous. Numbers form the basis on which to establish whether someone meets crucial criteria and gets included in the pool of those waiting for transplantation or not. Among all these numbers, the pivotal one for a liver transplant, though, has been the MELD (Model of End-Stage Liver Disease) score since its adoption from the United States in 2006. This score, a number between six and forty that almost every patient I met knew by heart, functions as the crucial number in liver transplant medicine. It serves as a signpost for one's chances of a transplant, as the higher the number the better one's chances. This score has real-life consequences for patients like Arthur whose data is being calculated. After being registered in Eurotransplant's

---

[4]This variation in relative numbers can be explained due to the different legal frameworks for organ donation in these countries. Unlike other member states, Germany follows an opt-in system, requiring people's active consent for organ donations. As numbers have remained low despite major campaigns promoting organ donation, a change to an opt-out system is currently being promoted by the German health minister and is causing heated debates. As transplant medicine relies on the use of human body parts and tissues, the question of organ donation remains a contested and ethically charged issue. Especially so in a country that has a history of bodies being misused for experiments under National Socialism, or bodies being considered state property in the East German socialist state (Hogle 1999, 3).

data pool, and thus 'listed' for a transplant, patients' engagement with the ways in which their livers, and lives, are assessed, evaluated and classified becomes more refined and their focus on 'getting on the waiting list' is replaced by concerns about their MELD score. What patients rarely talked about was the fact that the score actually assessed their mortality risk, calculating the probability of death without a liver transplant over the next 3 months.

The so-called lab MELD score is the result of a calculation that takes three laboratory variables into account, combining serum creatinine, serum bilirubin and the international normalized ratio for prothrombin time (INR) to calculate people's mortality risk. 'The original mathematical formula for MELD is: MELD = 9.57 × loge(creatinine) + 3.78 × Loge(total bilirubin) + 11.2 × Loge(INR) + 6.43' (Kamath and Kim 2007, 797). But not all end-stage liver diseases are considered with these clearly defined laboratory values. So, for instance, in the case of liver cancer patients like Arthur, the responsible transplant committee ascribes a patient a MELD that is supposed to 'match' their mortality risk as long as their cancer stays within clearly defined criteria regarding number and size of tumours, a so-called matchMELD (Bundesärztekammer 2017). The matchMELD is intended to make the comparison between patients possible in order to assess who is most urgently in need of a transplant—despite the wide range of ways in which liver diseases present themselves, as well as the singularity of each patient's experience. These calculations are driven by the effort to assess mortality risk and control probable futures, to make the unknown predictable and manageable (Rose 2001).

Like waiting lists that are in constant flux, the MELD score is also a temporary snapshot, a number that increases and decreases with patients' fluctuating physical conditions and changing medical values. A patient's MELD score, their calculated mortality risk, not only increases with their deteriorating health, also has the potential to drop when, due to the liver's capacity to regenerate healthy tissue, a person's physical condition improves. Because the score is grounded in medical values assessed in laboratories by scientifically proven biomedical methods, it represents the dominant ideal of evidence-based medicine and is surrounded by a sense of objectivity and impartiality. Information infrastructures and technologies like scores, lists and algorithms clarify such aspects as patients' health, mortality risk or chance of a transplant. But while they create the impression that the decision-making processes involved are conducted fairly and objectively, these medical information technologies obscure how these are shaped by political and moral economies, intimate considerations and individual decisions (Amelang 2014; Rehsmann 2018).

The MELD score is in line with the dominant trend of the increasing standardization of medical care. Most physicians I encountered in transplant clinics would have favoured a more individualized approach to medical care and criticized this trend as well as the emphasis on evidence-based medicine. Simultaneously, many of them also felt ambivalent about more individualized medical practice. Bureaucratic paperwork was already a big part of their daily work, and while standardized forms like checklists do have their limitations (Kocman et al. 2018), they also reduce the amount of documentation required from medics. Decisions about medical treatment derived from a less standardized approach would require much more detailed documentation of all decisions involved. What is more, they would also require more time—one of the most valuable resources in healthcare institutions and a critical one when treating people with failing livers.

Waiting-in-Uncertainty, Mobile Phones and Immobilities

Arthur Berger was one of many patients I came to know in the space of the transplant clinic who was looking for treatment for their liver diseases, getting assessed for a transplant or who were already on the wait list and, thus, in the very midst of the unknowns and uncertainties of waiting for a liver transplant. Slightly older than most other patients, who were usually in their 50s or early 60s, the 70-year-old was affected by irreversible liver cirrhosis as well as nonmetastatic liver cancer and without a transplant his prognosis was grim. Because no long-term substitute treatments are available for failing livers, like dialysis for kidney failure, receiving a donor liver becomes a highly urgent matter for those with irreversible liver failure. In the context of failing livers, time becomes an extremely limited resource—and waiting for a transplant a particularly trying experience. While Arthur was listed for a transplant, his eligibility for a donor liver was continuously reassessed by medical tests and the clinics' transplant board. For patients with liver cancer, the number as well as the size of tumours in the liver have to fit certain criteria. If their cancerous liver tumours exceed these strict limits, patients are no longer considered eligible for the treatment that they so urgently need, as they are assessed to have less chance of surviving the transplantation.

Thus, getting on the waiting list is not a unilinear process that cannot be revoked. Because of their fluctuating medical condition, patients are kept under continuous observation to assess their MELD score, to check their compliance regarding alcohol consumption or to ascertain whether their liver tumours fit the eligibility criteria. In other words, Arthur's eligibility for treatment and with it his chance of survival were not the result of a one-off assessment. Getting access to the waiting list was not the end of medical tests, treatments and assessments. What followed was a continual process of evaluation that lasted for years, with the constant threat of losing his eligibility.

In addition to the same process that all patients must go through, regardless of their age or condition, Arthur's eligibility was further scrutinized due to his advanced age—despite the fact that there is no official age limit for transplant recipients. It was his treating physician who told me that he had argued in Arthur's favour based on his biological age, because his physical constitution was that of a 60-year-old. Providing that the growth of his tumours was kept at bay, Arthur's eligibility for a liver transplant, i.e. his access to the waiting list, was granted for one more year. His doctor told me that if he did not receive a donor organ in that set timeframe, he would be considered too old and no longer eligible. So, without knowing how much time he had left without a transplant and how long he would have to wait for a donor liver, Arthur's chances got slimmer and slimmer with every month that passed by and he became increasingly impatient due to his medical emergency. His eligibility was as fragile as his health and while Arthur was glad that he was on the waiting list for a transplant, he was concerned whether he would outlive his wait.

When I visited Arthur at home, his mobile phone was omnipresent. It was always by his side—he took it with him when he went into the kitchen, into the bathroom—and he could not bear not having it next to him. He told me that since he had got on the waiting list, he always kept his phone within reach and made sure that it was charged all the time and indeed switched on because missing a call, *the* call, could mean missing out on his chance of survival. As a tangible manifestation of the invisible waiting list, the phone seemed to have become his constant companion—in his daily life and in his dreams. The desire to hear his phone ring had become so strong for Arthur that he repeatedly 'heard' it ringing—mistakenly, as it turned out, again and again—and that the wished-for ringing sound even followed him into his dreams.

The mobile phone not only symbolized Arthur's hope for a transplant was also a continual reminder of his dependency on medical expertise and technologies. His imaginings of the transplant surgery followed him into the recurrent nightmares that he suffered from. Again and again, he was lying conscious and cut open on the operating table, with surgeons removing flesh and surreal objects such as a children's bicycle from his open abdomen. Again and again, he woke up in a cold sweat in the middle of the night from nightmares so unsettling that he became afraid of falling asleep. While Arthur's dream about the transplant clinic calling him on top of a mountain with his chance of a transplant was significantly less violent than these surgical nightmares, it also exemplified his unease with his life depending on a donor liver, and the sense of 'existential immobility' and 'stuckedness' he experienced while 'waiting out' (Hage 2009) his wait. With his survival being dependent on a transplant and his lack of control over the duration and outcome of his wait, Arthur felt like he had no choice but to endure this wait. Unable to live life the way he wanted, he felt stuck and suffered from the uncertainties this waiting period entailed. Nonetheless, he kept on waiting, as 'the more one waits and invests in waiting, the more reluctant one is to stop waiting' (ibid., 104), and Arthur did not want to miss out on his chance of a liver.

Glancing again and again at the phone, lying silent on the table, Arthur told me that he was bothered by the fact that he could not travel at the moment. After all, more travelling was one of the main reasons why he wanted a transplant. 'I'm not done yet', he said, 'I want to see so much more of the world, have more time with my wife and see my grandchildren grow up'. Arthur's mobile phone not only symbolized the possibility of treatment and the hope of prolonging his life, but was also an expression of his current immobility. If he planned to travel, he had to inform his transplant clinics, and he would temporarily be considered 'nontransplantable' for the time of his travel. He was still free to travel, but what if a donated organ became available during exactly that time? 'I couldn't bear to potentially miss out on that', Arthur said. Thus, he preferred to stay close to the clinic; 'day trips are fine, but all other trips have to wait for now'. Somehow, ironically, the *mobile* phone, which Arthur carried everywhere he went, seemed to make him *im*mobile, as he had to stay close to his transplant clinic. Not being able to move as he wanted to made him feel 'like a chained yard dog with a collar around my neck', restricted in his movements and his freedom.

I want to argue that the particular workings of the 'hope-generating machine' (Nuijten 2003) of transplant medicine, with its intangible, fleeting waiting lists and its invisibly calculating algorithms, give rise to a waiting that is characterized not only by uncertainties but also by ambivalences of mobility and immobility. The mobile phone is in itself a symbol of mobility. But in the context of waiting for a liver transplant, the mobile phone symbolizes several other things, as the philosopher Francisco J. Varela (2001) pointed out in a phenomenological reflection on his liver transplantation. The procedures of being assessed and evaluated for transplantation were only the beginning of a time in which the mobile phone would become ever present:

> After months I was requested to carry on me at all times a dedicated portable phone, and to never be far from the hospital. [...] Weeks without end; every minute the pressure of my portable phone as witness awakening me to the immense fragility of my life and the tenousness [sic] of my identity in this tangle of deferred causalities. (Varela 2001, 267)

The mental strain involved in being reachable day and night was also articulated by other patients, such as Erika Schneider, a woman in her 60s who was waiting for a donor liver around the same time as Arthur:

'This waiting definitely puts you in a state of restlessness. Because you always have to be reachable by telephone—even at night'. Even after receiving a transplant, the emotional stress of being 'always ready' for 'the call' remains a defining experience in patients' accounts. Florian Weiss, a 24-year-old who had received a liver about a year before we met, recalled his time of waiting for his transplant: 'You've got to be ready anytime; have your bag packed and the phone always in reach'. A psychologist I interviewed compared this state of restlessness and of always being-ready with the last 4 weeks of pregnancy, when expectant parents have to be ready anytime with their bags packed for the clinic. However, in the context of liver transplants this time is not limited to 4 weeks but can last indefinitely for weeks, months or even years.

It becomes clear that in the context of waiting for a liver transplant, the mobile phone is more than just a technological communication tool that makes it easier for clinics to reach patients. While the mobile phone signifies the ever-present hope for and chances of a transplant, it is also an uncomfortable reminder of one's 'fragility' (Varela 2001, 267) and 'stuckedness' (Hage 2009) in waiting for a liver. Moreover, the 'request' (Varela 2001, 267) to be reachable is often experienced as a duty that causes an additional emotional burden during waiting. Thus, while carrying the hope for a future, the mobile phone also puts a strain on the relation between the patient who is waiting and the phone that might deliver the call. Hence, I argue that the mobile phone shifts back and forth between being a mere technological communication tool and a symbol of future aspirations, medical contingencies and existential im/mobilities of those who are waiting for a liver.

While landlines are clearly defined in their radius of reachability, mobile phones offer the possibility of being accessible everywhere at any time, allowing increased flexibility and mobility for those waiting for their transplant clinic to call. To receive the phone call from the transplant clinic carries the potential to offer a hoped-for ending to people's waiting time and 'existential and physical immobility' (Hage 2009). The mobile phone is a reminder that the sense of 'stuckedness' that is experienced during this 'expectant waiting' (Elliot 2016) for a donor liver might lift at any time, allowing life to move on.

Conclusion

In this chapter, I have examined the workings of transplant medicine as a 'hope-generating machine' (Nuijten 2003) by scrutinizing such critical features as waiting lists, allocating algorithms and mobile phones. I have discussed the waiting list as a crucial step for patients towards fulfilling their hope for life-saving treatment and delineated its differences from more common understandings of lists as fixed order. While lists seem to be a reasonable administrative tool to distribute goods among people, I have argued that waiting lists for liver transplants possess particularities that are better understood as articulation of transplant medicine's invisible digital infrastructure, especially due to the importance of intangible algorithms in creating these fleeting lists. By taking a closer look at the particularities of the exemplary waiting experience of Arthur Berger, I have illustrated how the invisible workings of the hope-generating transplant machinery affect those whose livers and lives are continuously assessed and evaluated while waiting for a liver. During this uncertain waiting time, patients orient their hopes according to these fleeting lists and the invisible calculations informing them. The fluctuating figure of the MELD score, which represents one's mortality risk and chance of a transplant, gives patients some indication of their hope for a donor liver and becomes a fluid benchmark symbolizing their chances of death *and* survival.

Liver transplant medicine is a field in which vital resources are too scarce to treat every person in need, making the death of some inevitable. Triaging livers and lives is an essential aspect of the field, in which the allocation of donor livers is delegated to waiting lists that are based on computer algorithms. With the invisible workings of these algorithms, patients' chances of survival are calculated as part of a complex digital infrastructure that generates temporary waiting lists for individual donor organs. Due to their complexity and fleeting characteristics, the powerful workings of this invisible infrastructure are hard to grasp for those waiting for a transplant and affected by it. It comes as no surprise that, instead, people talked about *the* waiting list rather than about multiple fleeting lists, algorithms, data pools and computer programs.

Continuous assessment and evaluation of livers and lives by means of medical tests controls access to this temporary data pool. Hence, while waiting lists are an allocation tool, a bureaucratic technology of waiting and a legally necessary step on the way to a transplant, they also serve as a marker of eligibility and symbolize one's chances of a liver transplant. By carrying the possibility of a life-saving liver transplant, the waiting list fuels patients' hope for a future, offering a canvas for their hopes, fears and expectations. Arthur's experience illustrates how the hope for survival, and for existential and physical mobility, informs patients' decision to wait for a transplant, while this waiting then causes a sense of 'stuckedness' and generates physical and existential 'immobilities' (Hage 2009). It is the aspiration for mobility and a future that maintains patients' hope for a transplant and leads to 'expectant waiting' (Elliot 2016) that is characterized by uncertainties regarding its outcome and indeterminacy regarding its duration. I have argued that during this particular temporal experience of waiting, the mobile phone becomes an essential, but ambivalent part of the infrastructure of the 'hope-generating machine' of transplant medicine. During patients' wait for a transplant, the mobile phone, as a symbol of mobility, transforms from a mere communication tool to an ambivalent marker of people's simultaneous mobility and immobility. In this time of uncertain waiting, the mobile phone becomes an extension and tangible manifestation of the fleeting and invisible waiting list, a torturous reminder of one's fragility and dependency on medical expertise, while simultaneously always carrying the potential to mark the end to this uncertain wait.

Acknowledgements The research was funded by the Swiss National Science Foundation (project number 149368 and 175223). I would like to thank the "Intimate Uncertainties" research team, Sabine Strasser, Veronika Siegl, Gerhild Perl and Luisa Piart, as well as Janina Kehr who commented on previous versions of this chapter. Special thanks to all who shared their valuable time and experiences while waiting and hoping for a liver.

References

Amelang, Katrin. 2014. *Transplantierte Alltage: Zur Produktion von Normalität nach einer Organtransplantation.* Bielefeld: Transcript Verlag.

Auyero, Javier. 2012. *Patients of the State: The Politics of Waiting in Argentina.* Durham and London: Duke University Press.

Bandak, Andreas, and Manpreet K. Janeja. 2018. "Introduction: Worth the Wait." In *Ethnographies of Waiting: Doubt, Hope and Uncertainty*, edited by M. K. Janeja and A. Bandak, pp. 1–39. London, Oxford, and New York: Bloomsbury Academic.

Bowker, Geoffrey C., and Susan Leigh Star. 2000. *Sorting Things Out: Classification and Its Consequences.* Cambridge, MA and London, UK: MIT Press.

Bundesärztekammer. 2017. Richtlinien zur Organtransplantation gem. § 16 TPG: Richtlinie gemäß § 16 Abs. 1 S. 1 Nrn. 2 u. 5 TPG für die Wartelistenführung und Organvermittlung zur Lebertransplantation. In *Deutsches Ärzteblatt*. Bundesärztekammer, ed. pp. 1–20, Vol. 114.

Connolly, Kate. 2013. "Mass Donor Organ Fraud Shakes Germany." *The Guardian*, January 9. Accessed May 18. https://www.theguardian.com/world/2013/jan/09/mass-donor-organ-fraud-germany.

Crowley-Matoka, Megan. 2016. *Domesticating Organ Transplant: Familial Sacrifice and National Aspiration in Mexico.* Durham, NC and London: Duke University Press.

Day, Sophie. 2016. "Waiting and the Architecture of Care." In *Living and Dying in the Contemporary World: A Compendium*, edited by V. Das and C. Han, pp. 67–184. Oakland: University of California Press.

Ehn, Billy, and Orvar Löfgren. 2010. *The Secret World of Doing Nothing.* Berkeley, Los Angeles, and London: University of California Press.

Elliot, Alice. 2016. "Paused Subjects: Waiting for Migration in North Africa." *Time & Society* 25 (1): 102–116.

Eurotransplant. 2015. *Eurotransplant—More Than a Match.* Accessed April 19, 2019. https://vimeo.com/124284017.

Goody, Jack. 1977. *The Domestication of the Savage Mind.* Cambridge, London, New York, and Melbourne: Cambridge University Press.

Graeber, David. 2015. *The Utopia of Rules: On Technology, Stupidity, and the Secret Joys of Bureaucracy.* Brooklyn and London: Melville House Publishing.

Hage, Ghassan. 2009. "Waiting Out the Crisis: On Stuckedness and Governmentality." In *Waiting*, edited by G. Hage, pp. 97–106. Carlton, VIC: Melbourne University Press.

Hogle, Linda F. 1999. *Recovering the Nation's Body: Cultural Memory, Medicine, and the Politics of Redemption*. New Brunswick, NJ and London: Rutgers University Press.

Kamath, Patrick. S., and W. Ray Kim. 2007. The Model for End-Stage Liver Disease (MELD). *Hepatology* 45 (3): 797–805.

Kaufman, Sharon R., and Lakshmi Fjord. 2011. "Medicare, Ethics, and Reflexive Longevity: Governing Time and Treatment in an Aging Society." *Medical Anthropology Quarterly* 25 (2): 209–231.

Kierans, Ciara, and Jessie Cooper. 2011. "Organ Donation, Genetics, Race and Culture: The Making of a Medical Problem." *Anthropology Today* 27 (6): 11–14.

Kocman, David, et al. 2018. "Neither Magic Bullet Nor a Mere Tool: Negotiating Multiple Logics of the Checklist in Healthcare Quality Improvement." *Sociology of Health & Illness* 47 (4): 1–17.

Nuijten, Monique. 2003. *Power, Community and the State: The Political Anthropology of Organisation in Mexico*. London and Sterling, VA: Pluto Press.

Porter, Theodore M. 1996. *Trust in Numbers: The Pursuit of Objectivity in Science and Public Life*. Princeton, NJ: Princeton University Press.

Reed, Adam. 2011. "Hope on Remand." *Journal of the Royal Anthropological Institute* 17 (3): 527–544.

Rehsmann, Julia. 2018. "Confined Live(r)s: Self-Infliction and Arbitrary Survival in the German Transplant System." *Anthropological Journal of European Cultures* 27 (2): 45–64.

Rose, Nikolas. 2001. "The Politics of Life Itself." *Theory, Culture & Society* 18 (6): 1–30.

Sharp, Lesley A. 2014. *The Transplant Imaginary: Mechanical Hearts, Animal Parts, and Moral Thinking in Highly Experimental Science*. Berkeley, Los Angeles, and London: University of California Press.

Shaw, David. 2013. "Lessons from the German Organ Donation Scandal." *Journal of the Intensive Care Society* 14 (3): 200–201.

Strasser, Sabine, and Luisa Piart. 2018. "Intimate Uncertainties: Ethnographic Explorations of Moral Economies across Europe." *Anthropological Journal of European Cultures* 27 (2): v–xv.

Varela, Francisco J. 2001. "Intimate Distances: Fragments for a Phenomenology of Organ Transplantation." *Journal of Consciousness Studies* 8 (5–7): 259–271.

Open Access This chapter is licensed under the terms of the Creative Commons Attribution 4.0 International License (http://creativecommons.org/licenses/by/4.0/), which permits use, sharing, adaptation, distribution and reproduction in any medium or format, as long as you give appropriate credit to the original author(s) and the source, provide a link to the Creative Commons license and indicate if changes were made.

The images or other third party material in this chapter are included in the chapter's Creative Commons license, unless indicated otherwise in a credit line to the material. If material is not included in the chapter's Creative Commons license and your intended use is not permitted by statutory regulation or exceeds the permitted use, you will need to obtain permission directly from the copyright holder.

3

'Being Stuck': Refugees' Experiences of Enforced Waiting in Greece

Pia Juul Bjertrup, Jihane Ben Farhat, Malika Bouhenia, Michaël Neuman, Philippe Mayaud, and Karl Blanchet

What is meant for us? We are stuck now, we do not know anything. Will our story be that we will stay in Europe, or will we be forced back to Afghanistan? Will we be able to develop as individuals? These are the questions we ask ourselves.

—Afghan woman, Greece, November 2016

P. J. Bjertrup (✉) · M. Bouhenia
Epicentre, Médecins Sans Frontières, Paris, France
e-mail: pjb@anthro.ku.dk

P. J. Bjertrup
Department of Anthropology, University of Copenhagen, Copenhagen, Denmark

J. Ben Farhat
Intervention Epidemiology and Training, Epicentre, Médecins Sans Frontières, Paris, France

Introduction

In March 2016, the border of the Former Yugoslav Republic of Macedonia (FYROM) closed and around 60,000 refugees became stuck in Greece while waiting for a resolution on their asylum claims. The closure of the FYROM border was one of several initiatives that attempted to halt movement of refugees and migrants into the European Union (EU). By 2015, 1,032,408 refugees and migrants arrived in Europe and 3771 lives had been recorded lost during the journey (UNHCR 2019). Images of dinghies and orange lifejackets that covered entire beach areas became daily occurrences as the media covered the 'migration' or 'refugee' crisis unfolding in Europe.

While Greece had been an entry and transit country into other European countries for several years, the border closure in 2016 forced the country to adapt to hosting and processing large numbers of asylum claims. The conditions and processing time for these claims have been criticized by several international non-governmental and human rights organisations as causing or worsening mental health distress among the refugee population (Human Rights Watch 2017; Amnesty International 2016; MSF 2017).

This chapter narrates the ways refugees in Greece were rendered immobile and their different modes and experiences of waiting. Waiting has traditionally been considered a passive state, and Ghassan Hage (2009a) identified a sense of 'stuckedness' as the cornerstones of social life become suspended when liminality turns into protracted waiting

---

M. Neuman
CRASH, Médecins Sans Frontières, Paris, France

P. Mayaud
Clinical Research Department, Faculty of Infectious and Tropical Diseases, London School of Hygiene & Tropical Medicine, London, UK

K. Blanchet
Geneva Centre of Humanitarian Studies, University of Geneva, Graduate Institute, Geneva, Switzerland

(Hage 2009a). The conflation of waiting and passivity has been challenged by disclosing various degrees of activity within the wait (Bendixen and Eriksen 2018; Brun 2015; Rotter 2016). Furthermore, 'waiting is both a technique of disciplining and a phenomenological event' (Haas 2017, 78), and the two facets of waiting are intertwined. In this chapter, we mainly engage with waiting as a phenomenological event and how it was experienced by refugees in Greece. However, we also attempt to show how refugee camps and arcane and bureaucratic asylum processes are techniques of disciplining the refugees, and how the asylum system offers possibilities and hope in different ways and degrees, making some refugees more likely to break out of the open-ended waiting, while others endure the wait.

Political Context and European Border Closures

More than lines on a map, borders are employed to govern populations, facilitating some forms of mobility while restricting others, thus creating an 'uneven landscape of movement' (Castañeda 2018). Among others, this is exemplified by the illegalisation of refugee and migrant movement between borders. During 2016, increased border control measures and a cascade of border closures were implemented to restrict the mobility of refugees and migrants within Europe. In addition, the European Commission reached an agreement with Turkey, the so-called EU-Turkey deal, which specified that individuals crossing illegally from Turkey to Greece after 20 March 2016 should be returned to Turkey if they were not applying for, or were not eligible for, asylum (The European Council 2016). The deal specified that for every person of Syrian nationality deported from Greek islands to Turkey, the EU would offer asylum to one Syrian living in a Turkish refugee camp. By March 2016, 60,000 refugees were rendered immobile in an enforced waiting in Greece of which around 8500 were held on the Greek islands and denied travel to the mainland. In Greece, the refugees arriving were mainly from Syria (47%), Afghanistan (24%) and Iraq (15%) (UNHCR 2016).

The arrival of Syrian refugees in Europe caused a redefinition of the asylum system. Their rightful claim to asylum could not be countered, however, discourses shifted from the 'rights of refugees' to 'who is a genuine refugee' (Dimitriadi 2018, 2). After the EU-Turkey deal, refugees in Greece were subjected to different asylum rules and processes depending on nationality and time of arrival in Greece. Those who arrived after the EU-Turkey deal were required to remain in Greek island facilities until the full registration of their asylum claim. Certain nationalities (e.g. Syrians, Eritreans and Iraqis) on the mainland and those arriving before 20 March had the possibility to apply for 'relocation', which is the term used for the transfer of asylum seekers who are in clear need of international protection from one EU Member State to another. Afghans, the second largest refugee population in Greece, could not apply for relocation.

Methodology

This chapter is based on a study of refugee mental health and experience of violence that was commissioned by Médecins Sans Frontières (MSF) France to inform their operations in Greece. The study was conducted in collaboration with Epicentre and the London School of Hygiene and Tropical Medicine between November 2016 and February 2017 and consisted of an epidemiological survey and an explanatory qualitative study in refugee camps across Greece. The latter study is the focus of this chapter. The methodology of the overall study has been published elsewhere (Bjertrup et al. 2018; Ben Farhat et al. 2018).

In brief, the qualitative study took place in four refugee camps on mainland Greece (in the regions of Attica and Epirus), in a squatted building in the centre of Athens, and at a reception and identification centre on the island of Samos, the so-called Samos Hotspot. Adult refugees were recruited for in-depth interviews (IDIs) through the epidemiological survey conducted in parallel. Forty-seven IDIs, thirty-eight individual interviews and nine paired-interviews with spouses or relatives were conducted, with a total of fifty-six participants (twenty-five women and thirty-one men). The interview participants included nine different nationalities with a majority of Syrians. Herein, we present results that emerged during interviews in relation to themes of uncertainty, waiting and hope. The interviews lasted between one to two hours and were conducted in an MSF on-site mental health clinic, in the shelter of the person being interviewed (container, tent or hotel room) or in the back of the research team's outreach vehicle. The first author conducted the interviews, with the assistance of Arabic, Farsi, and Kurdish (Kurmanji) interpreters. Participants were informed that their decision to participate in the study or not would have no implications in terms of access to health services or on their asylum case. During the interview, participants were reminded of their voluntary participation, their right not to answer questions and to terminate the interview at any point. Lastly, participants were also informed about the offer of free on-site mental health services provided by MSF. The IDIs were supplemented with five focus group discussions, which explored everyday life opportunities, problems and challenges faced by the refugee community at the different sites, for example, asylum procedures, work, education and social activities. To protect participants' identities, all individuals named in this chapter have been assigned pseudonyms.

Waiting in Refugee Camps—Passivity and Dependency

Most camps on mainland Greece accommodated refugees of the same nationalities (with few exceptions), while the reception centre on Samos island included a variety of nationalities. The Samos centre was designed as a closed camp and even though it did not function as such, the fence and barbwire surrounding the camp conveyed a feeling of detention. The centre was overcrowded, housing more than 1100 people with a capacity of 700; it lacked proper shelter and basic water and sanitation facilities. In central mainland Greece, the refugee camp sheltering mainly an Afghan population lacked proper sheltering facilities, the presence of non-governmental organizations was poor and generally the camp's living conditions were rudimentary compared to a camp just 40 km away hosting Syrian refugees. Until December 2016, when they were slowly moved into containers, the Afghan population lived in small pop-up tents.

The organization, physical structures and location of the refugee camps fostered passivity and submission to a large extent. The camps were generally located in isolated places and the image portrayed by Michel Agier (2008) of the refugee camp as a 'place of waiting apart from society' (Agier 2008, 40) seems fitting. In Greece, life in the camps was structured around queuing for everyday necessities such as sanitary facilities, food, health and administration services, which reinforced a feeling of physical immobility:

> Usually, I start my day with standing in the food line, and then after breakfast I sometimes queue at the tea house to charge my phone. Other days, I go down town to work and to buy stuff that we need. But it is difficult, because we don't have a work permit and the police stops us. (Afghan man, November 2016)

In most camps, there was not much to do, time passed by slowly and many refugees spoke of their life as only consisting of sleeping and eating. During interviews, some participants raised concerns about relatives who seemed depressed, staying in bed and inside all day, only leaving the shelter when absolutely necessary. For most refugees, the experience in Greece was difficult and something they had to 'wait out' (Hage 2009b). At times, participants expressed feelings of being treated like animals, when talking about conditions in camps and an Afghan woman uttered: 'We are just led to barely survive'. Describing life as limited to 'bare survival' points out how life in a refugee camp was, in many ways, experienced as 'social death'. Similar to this, Agier (2008) notes: 'The application of the [humanitarian] principle established a contradiction between minimal biological life (protection, feeding, health) and the social and political existence of individuals: the refugees are certainly alive, but they no longer "exist"' (Agier 2008, 49).

In the camps, refugees were assisted and depended on external aid relief, which included a monthly cash transfer, food, shelter and second-hand clothes, all of which created a sense of dependence. Being dependent had negative connotations and negatively affected self-esteem:

> I was really humiliated here [in Greece]. I lived through the bombings in Syria, there at least I had dignity. Here, I lost it completely. When you have to stay for a long time in line just to bring food to your children. My son asks me for some chips but I can't buy it for him. In Syria, I was buying everything, clothes, food, everything. (Syrian man, November 2016)

Refugees were expected to queue for everyday necessities and hand-outs and wait for the long bureaucratic processing of their asylum claims. Most refugees experienced queuing, receiving hand-outs, and waiting in camps as demeaning, fostering passivity and dependence. Some refugees recounted how they felt uncomfortable and embarrassed when fellow refugees would make problems in the camps and act in a manner that they were not supposed to:

> Here some of the refugees are really stupid. They make a problem about a pair of shoes and food, so you feel really embarrassed. It is for all of us, why do they behave like this? (Syrian man, November 2016)

For Hage (2009b), 'waiting out' involves both a subjection to certain social conditions, a braving of these conditions and it is a 'governmental tool that encourages a mode of restraint, self-control and self-government' (Hage 2009b, 102). Auyero (2011, 2014) argues how: 'the waiting room is an area of compliance' (2011, 21) for poor people waiting in welfare offices in Argentina. In a similar vein, compliance is embodied among the refugees, in that most queue for everyday necessities, they wait for an answer on their asylum claim and they most often do so quietly and patiently. The remark by the Syrian man on feelings of discomfort and embarrassment when other refugees 'made problems' as if this 'bad behaviour' reflected back on him illustrates how a form of compliance or self-disciplining is embodied.

Uncertainty and Suffering Linked to the Asylum System

> How long are we going to sit and wait? We are going insane of having nothing to do. We are not going forward and we do not want to look back. (Afghan woman, November 2016)

The quoted Afghan woman excellently expressed how life for most refugees was felt: a standstill where the next step remained unknown both in terms of what this next step would constitute, and when this would happen. The overall experience in Greece was one of waiting for life to resume pending a resolution on the asylum case. There was among the refugees a strong sense of missing out on life and of wasting or losing time. In Hage's (2009b) terms, there was a sense of stuckedness or existential immobility, that one was not going anywhere in life, especially in the sense that children were missing out on school and on the chance of having a good future. In Greece, the experience of uncertainty, obscurity, and lack of control over the asylum processes, procedures and outcome largely pervaded the refugees' narratives:

> On the island they told us: 'in two months you will get your reunification' and till now there is nothing, they forgot us. The procedures are very long, it takes around seven months between interviews. They report the problem and delay it. They didn't give us the correct information, they give us false hope, but you have to adapt to everything, because you don't have a choice. If you want to arrive at your destination, you have to be patient. (Syrian woman, December 2016)

The asylum procedures and processes were unclear to most refugees and limited information was provided by the authorities, thus creating a deep sense of uncertainty. Some refugees did not know if they were in an asylum process at all, while others were unsure if they were in a process of relocation or family reunification.

Refugees often described states such as stress, sadness, loneliness, depression, worrying and insomnia. While most refugees could recount experiences of war, violence and potential 'traumatic' events in their home country and/or during the flight, only a few related this to their current poor well-being. Instead, most located the source of their pain and suffering to the uncertainty of the present. In addition, the current moment and living conditions in camps in Greece were also often identified as the cause of poor well-being:

> [W]hen living in dirty conditions, mental health issues arise. If they took us out from this dirty, uncomfortable place, mental health issues would be resolved on their own. Everybody will recover without the help of a doctor [psychologist]. [We] do not have it well here, the food, the precarious living conditions, this all contributes negatively to our mental health. We just want to go on, live in a good place, in a good home. Then all these negative thoughts will disappear. (Iraqi man, December 2016)

The refugees were often consumed with anxious thoughts about what they lacked: a legal status, security, a home, a job, their families and an education for their children. The refugees' experiences of life disruption, or lives 'on hold', meant that the present was often 'hyper-realized' (Haas 2017). Instead of the present receding into the background, the refugees' lives were characterized by an 'enforced orientation to the present' (Genova 2002 in Haas 2017, 82)—a present that was deeply painful for most, as it was characterized by uncertainty, absence of relatives, of a home, a job and so on. While uncertainty largely permeated almost all narratives, the degree of experienced uncertainty varied according to the specific asylum process and procedures that the refugees had the possibility to apply under. In the following parts, we show how the refugees experienced and navigated 'existential insecurity' and open-ended waiting.

'Existential Insecurity' and Open-Ended Waiting

We constantly heard references to, or descriptions of, 'being stuck', 'being in a prison' or 'being kidnapped' when refugees were describing their current life. The descriptions and embodied experiences denote the lack of control the refugees experienced; their life and asylum decision were in the hands of a powerful Other. On the topic of time and power, Bourdieu (2000) writes: 'Absolute power is the power to make oneself unpredictable and deny other people any reasonable anticipation, to place them in total uncertainty by offering no scope to their capacity to predict' (Bourdieu 2000, 228). This 'total uncertainty' and an absence of 'capacity to predict' were experienced by refugees in Greece where the asylum system offered no, or very little chance of being resettled in a European country. These refugees were potential 'citizens in waiting' and 'deportees-in-waiting' (Haas 2017), and this duality emerged from their narratives:

> What will and will not happen? Will we stay here, or be moved? Will they deport us back to Afghanistan? All these questions have a huge impact on our mental state. Many of us are struggling psychologically now. It will drive you crazy not knowing what will happen to you and your loved ones. (Man from Afghanistan)

As the quote clearly demonstrates, anxious thoughts about the future and the fear of deportation occupied the present life in Greece and could be characterized as an 'existential insecurity'. It was especially among the Afghan population and the refugee population on Samos Island that the fear of being deported occupied thoughts and was the cause of much worrying. The pressure on the asylum system was stronger on the islands than the mainland, which created further delays and postponements of interviews. As the Syrian population was initially the only nationality being processed on the islands, other nationalities faced significant delays and postponement of their admissibility interviews. Multiple postponements without explanation were seen by some as a sign of upcoming deportation.

Asylum processes and procedures in Greece were largely dependent on nationality and time of arrival and thus offered different possibilities and degrees of chance and hope for resettlement in Europe. Hage (2003) underlines the role of national societies in the generation and distribution of hope and highlights how hope is often distributed unequally between different groups of people (Hage 2003). Dominant notions of hope revolve around a sense that one is going somewhere in life, although a great proportion of the population experiences stuckedness (Hage 2009b, 97). Among refugees in Greece, dominant notions of hope revolved around settling in a European country and having a 'stable' and 'normal' life, signifying having a home, security, work, education for children and being with family members.

The EU asylum policies were largely favourable to people with Syrian nationality and a few other nationalities who had arrived in Greece before 20 March 2016, as they could apply under the relocation scheme. The Afghan population and the refugee populations on the Greek islands were not allowed to apply under the relocation scheme and their chances of receiving a favourable decision on their asylum claim were minor, which caused their waiting to be more insecure and open-ended. Hage (2009b) stresses how people consider physical mobility when they experience stuckedness that is a crisis in their sense of existential mobility. Among the refugees in Greece, the feeling of being stuck, the open-ended uncertainty and existential insecurity made some, especially those less likely to receive a favourable asylum decision, strongly consider movement with smugglers:

> I feel forced to send someone forward, because we are stuck here. The only thing left for us to do is to send some of us forward with smugglers. We do not have the money to go all together. The smugglers are raising the prices every day because they know we are desperate after being here for nine months. Every day, my husband and I are fighting over this. One day, he says I should go with our daughter, the next day it is him and our son. We do not want to be separated, but we can no longer be stuck either … We have been here for nine months. We are still alive, but the biggest issue is that our children are missing out on school. Time goes and they are left behind. (Afghan woman, November 2016)

For the different categories of refugees in Greece, the asylum system offered unequal chances and possibilities of resettlement in Europe, something that the refugees became acutely aware of. The Afghan population, and the population on the island of Samos in particular, experienced 'existential insecurity' and open-ended waiting, as the asylum system offered no or only a very slight chance of a positive outcome of their asylum claim. This made them more likely to risk going with smugglers.

Engaging with and Appreciating the Present Moment

While some refugees experienced an existential insecurity in relation to the present and the future, other refugees related more positively. One was Rasha, a 27-year old outgoing Syrian woman, whose waiting was mostly related to the present in an active and social way. We met Rasha and her daughters, aged five and seven, in November 2016 in a camp on mainland Greece, where they had been staying for almost 9 months. Rasha's husband was already in Sweden, and she had applied for family reunification. Being a teacher by profession and able to communicate well in English, Rasha volunteered with different organisations in the camp, offering translation and communication services between camp stakeholders and refugees. She also taught basic Arabic language skills to interested volunteers in the camp. She highlighted how being active was a way of distracting herself from the sorrow and worry:

> Sometimes, I miss my parents, my husband, my house. I miss all this but I also believe it is making me stronger … I want to forget my problems, so I go to work. I want to forget my story or all my pain, so sometimes I do sport, I go out to smoke shisha with friends … Sometimes you feel so happy just to have something to do. Before it was normal to smoke shisha all the time and it was not something important [in Syria], but now when you smoke shisha with your friend, it is very important. Just that, it makes you smile.

'Keeping busy' by socialising with volunteers and other refugees, volunteering in the camp, smoking shisha, playing with children, etc., was a tactic employed by many refugees that we met in Greece. This was, however, often challenging due to the lack of activities offered in camps and the lack of the right to work. For Rasha, her language skills, as well as her teaching background and general positive attitude, definitely played a role in her ability to use the waiting time in an active and social way. In addition, Rasha was settled in a camp that offered more opportunities and activities than other camps. Lastly, the fact that she qualified for family reunification probably facilitated a mode of waiting that was relational, active and meaningful. Rather than feeling the need to break out of frustrating waiting, Rasha was able to 'wait out' (Hage 2009b, 98). However, Rasha's engagement with the waiting process in Greece was more than just enduring or 'waiting out'. Small and seemingly unimportant activities such as 'smoking shisha' gained significance to her while in Greece and she was able to appreciate such activities in a new way. Similarly, other refugees described 'how drinking coffee and tea with others', 'going on trips' and 'going for a walk with children' made them feel good and like 'being at home' again. For Rasha, it seemed that the devastating war in Syria, the horrific flight with smugglers, and the difficult living conditions in Lebanon, Turkey and now Greece had reminded her to appreciate the 'little things' of the present moment. Other refugees expressed being 'thankful' for managing to find safety away from war and violence. Furthermore, Rasha talked about how all the hardship made her 'stronger', signifying a personal growth. Arriving in Greece was for her, as for many other refugees, a big disappointment and a great clash with what she had expected of a European country. When talking about hopes for the future, most refugees talked about settling in a European country and having a 'normal' or 'stable' life, meaning having a home, security, work, education for their children, and being together with family members. Rasha rarely expressed thoughts about the future, and when we explicitly asked her what her hopes were, she reluctantly answered: 'I cannot afford to have dreams now, I can only wish for basic things, a safe place to settle'. For Rasha and many other refugees, investing too much in the future could be potentially disappointing as she had learnt from her experience in Lebanon, Turkey and Greece, and attuning to and appreciating the 'now' was much safer.

Planning for an Anticipated Future

We met Zaher, a Syrian Engineer in his late forties, in December 2016, when he had already been in Greece for ten months. He had been away from his wife and son for more than two years, which caused him a great deal of emotional pain. Compared with Rasha's mode of waiting orientated towards the present, Zaher's mode of waiting was more orientated towards the future. By practising German every day and preparing for a future life when he would join his family in Germany, Zaher made the waiting in Greece more endurable:

> My second interview [for asylum] is in January. I am counting the days. I spent my time reading, learning German. At this moment, I am reading physics in German. I cannot understand the pronunciation, but as I was studying physics, I can understand the information, and I repeat until I understand. It makes me relaxed … My son speaks German very well, he went to school for four months and then he spoke German fluently. Every day, I talk with him about the job situation in Germany. I never expected to live in another country away from my wife and my son, and all the time it is difficult. To know he is successful encourages me to continue studying … I am optimistic. Maybe the future will be better.

At the beginning of his time in Greece, Zaher practised German language skills with a volunteer in the camp. However, at one point, volunteers were no longer allowed to come into the camp, and Zaher stopped practising with others. During the winter, he had had difficulties staying warm in his tent and he burned most of his books brought from Syria. Even though the conditions for keeping active and studying in the camp were far from ideal, Zaher continued reading the one book he had kept (on physics, in German) and continued repeating the pronunciation of the German words to himself. Zaher lamented that the volunteers were no longer allowed in the camp and he explained how he found the exchanges with the volunteers a good way of preparing himself for a 'new' life in Europe:

> Earlier, I suggested that the volunteers and the refugees meet and talk, not necessarily about politics, but for example about social life in Europe.

How can I best socialize in Europe, how should I approach people. This is a new life, how can I best do this? ... The Arab society is different than the European one, the customs, the habits, the behavior and the relationships between people ... To exchange such knowledge and experiences is a good idea. How will I integrate? I think that is a very good question. What are the needs of the European society and what are my needs? How can I participate in the society? We need to know such things.

Gasparini (1995) makes a distinction between those who wait for something indefinitely and those who expect something specific at a given time, with the latter being in greater control (Gasparini 1995). Active waiting implies anticipation and some confidence in the fact that a certain event will occur, and thus is closely connected to hope (Marcel 1967). Unlike Zaher, who did relate to the future with some confidence that he would eventually be reunited with his wife and son in Germany, most refugees whom we met in Greece did not show the same level of confidence and were thus either less able to prepare for the future or found it meaningless to do so. For most refugees, the future was difficult to anticipate, although the majority articulated hope in a future where they would have 'stability', often referring to the same kind of life they had led before having to flee. Zaher was eligible to apply for family reunification through the asylum system, and he was therefore more able to project himself into the future as he was waiting for something tangible. However, the waiting time for family reunification was long, making Zaher and refugees like him wait for an unknown amount of time.

Potentials for Mental Health and Psychosocial Interventions

Therapeutic approaches to refugee mental health primarily frame and seek to treat past trauma as the primary aetiology of suffering (Kienzler 2008). An understanding of the structure and organisation of refugee camps and of the asylum process as the primary sources of suffering, challenges such approaches and calls for more holistic mental health interventions. While the refugees in Greece did not deny past experiences of violence, war and suffering, they nonetheless saw the enforced waiting in refugee camps and the asylum system itself as responsible for both the continuation of pain and the production of new forms of deeply felt distress and suffering. In recent years, phased-based interventions to mental health among refugees, which emphasize provision of direct practical assistance and reinforcement of social networks, have gained ground (Rousseau et al. 2011; Inter-agency Standing Committee (IASC) 2007; UNHCR 2013). A recent study found that by targeting key issues for well-being, such as safety, rights and social networks through a range of practical and social activities, the mental well-being of participants was improved (Chase and Rousseau 2017). In a similar vein, we argue that in order to help refugees endure the wait and facilitate an active engagement, mental health and psychosocial interventions should recognize and draw on what refugees articulate as positively and negatively influencing well-being. Interventions could build on articulated meaningful activities that foster agency in waiting, e.g. social activities with other refugees and volunteers, language courses and volunteering in the camps. Furthermore, initiatives aimed at mitigating feelings of dependence (e.g. cash-transfers instead of ready-made and packed food) could also be introduced. Particular attention should be paid to each specific asylum context to ensure that initiatives and activities are relevant and meaningful. Language courses could be extremely relevant in situations of resettlement but could be seen as meaningless in a situation where deportation is likely. The above-mentioned recommendations, together with a fast and transparent asylum system, would help mitigate the distress experienced by refugees and could create a better foundation for mental health interventions in Greece and other places.

Conclusion

In Pascalian Meditations, Bourdieu (2000) wrote: '[...] one would need to catalogue, and analyse, all the behaviors associated with the exercise of power over other people's time both on the side of the powerful [...] and on the side of the "patient" [...]' (Bourdieu 2000, 228). In this chapter, we have attempted to respond to this plea. The different theoretic perspectives described in this chapter have, on the one hand, facilitated a subjective analysis of how waiting is experienced among refugees in Greece and have, on the other hand, enabled an analysis of the politics of waiting.

In Greece, refugee camps produce compliance (see Auyero 2011, 2014), and refugees have in many ways embodied the 'logics' of the refugee camp by queuing and waiting in line for food, sanitary facilities, health and administration services. This fosters feelings of passivity and dependence. Refugees associate their distress and suffering to the enforced waiting in the camps and especially to the tortuous asylum system. The present was deeply painful for most as it was characterized by uncertainty, lack of family members, a home, a job and so on. Uncertainty largely permeated almost all narratives. However, the degree of experienced uncertainty varied according to the specific asylum process and procedures that the refugees had the possibility to apply under. For Afghans and refugees on the Greek islands, the present was characterized as an 'existential insecurity', the anxiety and fear of deportation were strong as the asylum system offered no, or very little chance of being resettled in a European country. As a result, some refugees decided to break out of the frustrating open-ended waiting by going with smugglers instead of 'waiting out'.

While the waiting time for some refugees in Greece was inactive and passive, others managed to turn the empty or dead waiting time into something active and meaningful. Furthermore, others like Rasha were able to relate to the present in a different way where simple activities gained significance. Thus, instead of just 'waiting out', waiting in Greece was also an occasion to being thankful and appreciate the 'now' through small activities. The future was uncertain and difficult to anticipate for most of the refugees, although the majority articulated a hope in a future where they would have 'stability', signifying having a home, security, work, education for children and being with family members. Few were able to relate to the future with confidence and to plan for a future life in Europe as Zaher, who was in a process of family reunification, did.

Our study contributes to the body of research on immobility and mobility by taking, as a departure, the context of the refugee camp and waiting, a space and time normally considered temporary and immobile, which has been largely overlooked until the new mobility paradigm. We have shown how the refugees' waiting is both passive, dependent and uncertain on the one hand, and active, relational, and meaningful on the other. We have shed light on the affective dimensions of waiting by highlighting that the refugees' relation to the past, present and future is generating both distress and suffering, and meaning and hope. Acting towards hoped-for futures, some refugees break out of the open-ended waiting by going with smugglers while others, like Zaher, use the waiting time in Greece to prepare themselves for the future. In our study, mobility is not only a future-orientated but also a present-orientated modality. Through Rasha's case, we have shown that a waiting orientated towards the present, and an appreciation of the 'now', also offer the possibility for 'existential mobility' leading to self-growth and development. These affective dimensions of waiting and immobility, in general, could inform psychosocial interventions targeting refugee populations and other immobilized populations.

Acknowledgements We are incredibly grateful to the refugees who shared their stories and parts of their lives with us. Particular thanks go to our interpreters, who worked in difficult physical and emotional circumstances and to the dedicated MSF team in Greece for their support to the study. We would also like to thank the Editors and the anonymous reviewer for thoughtful and constructive feedback. Staff from the London School of Hygiene & Tropical Medicine wishes to thank the LSHTM Director and Faculty Deans who authorized and supported their participation in this research.

References

Agier, Michel. 2008. "Chapter 3: The Desert, the Camp, the City." In *On the Margins of the World: The Refugee Experience Today*, edited by Michel Agier and David Fernbach, English ed., 39–72. Cambridge, UK: Polity.

Amnesty International. 2016. "Greece: 'Our Hope Is Broken'—European Paralysis Leaves Thousands of Refugees Stranded in Greece." https://www.

amnesty.org/en/documents/document/?indexNumber=eur25%2f4843%2f2016&language=en.
Auyero, Javier. 2011. "Patients of the State: An Ethnographic Account of Poor People's Waiting." *Latin American Research Review* 46 (1): 5–29, 281, 285. http://dx.doi.org.ep.fjernadgang.kb.dk/10.1353/lar.2011.0014.
———. 2014. *Patients of the State, the Politics of Waiting in Argentina*. Durham: Duke University Press. http://ep.fjernadgang.kb.dk/login?url=https://ebookcentral.proquest.com/lib/kbdk/detail.action?docID=1173273.
Ben Farhat, Jihane, Karl Blanchet, Pia Juul Bjertrup, Apostolos Veizis, Clément Perrin, Rebecca M. Coulborn, Philippe Mayaud, and Sandra Cohuet. 2018. "Syrian Refugees in Greece: Experience with Violence, Mental Health Status, and Access to Information During the Journey and While in Greece." *BMC Medicine* 16 (March). https://doi.org/10.1186/s12916-018-1028-4.
Bendixen, Synnøve, and Thomas Hylland Eriksen. 2018. "Time and the Other: Waiting and Hope Among Irregular Migrants." In *Ethnographies of Waiting, Doubt, Hope and Uncertainty*, edited by Manpreet K. Janeja and Andreas Bandak. London: Bloomsbury Publishing PLC.
Bjertrup, Pia Juul, Malika Bouhenia, Philippe Mayaud, Clément Perrin, Jihane Ben Farhat, and Karl Blanchet. 2018. "A Life in Waiting: Refugees' Mental Health and Narratives of Social Suffering After European Union Border Closures in March 2016." *Social Science & Medicine* 215 (October): 53–60. https://doi.org/10.1016/j.socscimed.2018.08.040.
Bourdieu, Pierre. 2000. *Pascalian Meditations*. Somerset, UK: Polity Press. http://ebookcentral.proquest.com/lib/kbdk/detail.action?docID=5349073.
Brun, Cathrine. 2015. "Active Waiting and Changing Hopes: Toward a Time Perspective on Protracted Displacement." *Social Analysis* 59 (1): 19–37. https://doi.org/10.3167/sa.2015.590102.
Castañeda, Heide. 2018. "'Stuck in Motion' Simultaneous Mobility and Immobility in Migrant Healthcare Along the US-Mexico Border." In *Healthcare in Motion: (Im)Mobilities in Health Service Delivery and Access*, edited by Cecilia Vindrola-Padros, Ginger A. Johnson, and Anne E. Pfister. New York, NY: Berghahn Books.
Chase, Liana E., and Cecile Rousseau. 2017. "Ethnographic Case Study of a Community Day Center for Asylum Seekers as Early Stage Mental Health Intervention." *Journal of Orthopsychiatry*. https://doi.org/10.1037/ort0000266.
Dimitriadi, Angeliki. 2018. "Introduction: Delineating the Linkages." In *Irregular Afghan Migration to Europe: At the Margins, Looking In*, edited by

Angeliki Dimitriadi, 1–30. Migration, Diasporas and Citizenship. Cham: Springer. https://doi.org/10.1007/978-3-319-52959-2_1.

Gasparini, Giovanni. 1995. "On Waiting." *Time & Society* 4 (1): 29–45. https://doi.org/10.1177/0961463X95004001002.

Haas, Bridget M. 2017. "Citizens-in-Waiting, Deportees-in-Waiting: Power, Temporality, and Suffering in the U.S. Asylum System." *Ethos* 45 (1): 75–97. https://doi.org/10.1111/etho.12150.

Hage, Ghassan. 2003. *Against Paranoid Nationalism, Searching for Hope in a Shrinking Society*. Annandale, N.S.W: Pluto Press.

Hage, Ghassan. 2009a. *Waiting*. Melbourne: Melbourn University Press.

———. 2009b. "Waiting Out the Crisis: On Stuckedness and Governmentality." In *Waiting*, 97–106. Melbourne: Melbourne University Press.

Human Rights Watch. 2017. "EU/Greece: Asylum Seekers' Silent Mental Health Crisis Identify Those Most at Risk; Ensure Fair Hearings." https://www.hrw.org/news/2017/07/12/eu/greece-asylum-seekers-silent-mental-health-crisis.

Inter-agency Standing Committee (IASC). 2007. *Iasc Guidelines on Mental Health and Psychosocial Support in Emergency Settings*. Geneva: IASC.

Kienzler, Hanna. 2008. "Debating War-Trauma and Post-traumatic Stress Disorder (PTSD) in an Interdisciplinary Arena." *Social Science & Medicine* 67 (2): 218–227. https://doi.org/10.1016/j.socscimed.2008.03.030.

Marcel, Gabriel. 1967. "Desire and Hope." In *Readings in Existential Phenomenology*, edited by Nathaniel Lawrence and Daniel O'Connor. Englewood Cliffs, NJ: Prentice-Hall.

MSF. 2017. "Greece: Confronting the Mental Health Emergency on Samos and Lesvos—Why the Containment of Asylum Seekers on the Greek Islands Must End." http://www.msf.org/sites/msf.org/files/2017_10_mental_health_greece_report_final_low.pdf.

Rotter, Rebecca. 2016. "Waiting in the Asylum Determination Process: Just an Empty Interlude?" *Time & Society* 25 (1): 80–101. https://doi.org/10.1177/0961463X15613654.

Rousseau, C., K. Pottie, B. D. Thombs, M. Munoz, and T. Jurcik. 2011. "Appendix 11: Post traumatic Stress Disorder: Evidence Review for Newly Arriving Immigrants and Refugees." Canadian Medical Association. http://www.cmaj.ca/content/suppl/2010/06/07/cmaj.090313.DC1/imm-ptsd-11-at.pdf.

The European Council. 2016. "EU-Turkey Statement, 18 March 2016—Consilium." http://www.consilium.europa.eu/en/press/press-releases/2016/03/18-eu-turkey-statement/.

UNHCR. 2013. "Operational Guidance Mental Health & Psychosocial Support Programming for Refugee Operations." Geneva, Switzerland. http://www.unhcr.org/protection/health/525f94479/operational-guidance-mental-health-psychosocial-support-programming-refugee.html.

———. 2016. "Refugees & Migrants Sea Arrivals in Europe. Monthly Data Update: December 2016." https://data2.unhcr.org/ar/documents/download/53447.

———. 2019. "Situation Mediterranean Situation." https://data2.unhcr.org/en/situations/mediterranean.

4

'An (Im)Patient Population': Waiting Experiences of Transgender Patients at Healthcare Services in Buenos Aires

María Victoria Tiseyra, Santiago Morcillo, Julián Ortega, Mario Martín Pecheny, and Marine Gálvez

Introduction

Waiting time, considered as an indicator, has been used to measure the quality of health services. Up until now, waiting periods have been analysed mostly quantitatively in order to generate policies to reduce these

---

This article is a reworked version of: "(Im)Pacientes trans en hopitales públicos de Buenos Aires. La experiencia de la espera y la accesibilidad en contextos de estigmatización" *Vivência: Revista de Antropología de Universidade Federal do Rio Grande do Norte (Brasil)*, Vol. 1, No. 49, 2017.

M. V. Tiseyra (✉) · S. Morcillo · M. M. Pecheny
National Scientific and Technical Research Council (CONICET), Buenos Aires, Argentina

M. V. Tiseyra · S. Morcillo · J. Ortega · M. M. Pecheny
University of Buenos Aires (UBA), Buenos Aires, Argentina

Instituto de Investigaciones Gino Germani (IIGG), Buenos Aires, Argentina

M. Gálvez
Fullbright Scholar, Buenos Aires, Argentina

time periods (Siciliani et al. 2014; Aday and Andersen 1974). Quantitative analysis is certainly valuable, but we are interested here in a qualitative examination of waiting experiences as a way to approach the relationships of power, its dynamics and subjects implicated. We consider waiting as a social relationship that involves, at least, two subjects: one who is waiting, and the other, who is waited or produces that wait (Pecheny 2017). Thus, this chapter aims to examine how multiple structures of power intersect in waiting experiences and how they have a distinct significance for transgender patients.

In Argentina, the transgender community has endured a long history of violence, discrimination and exclusion, including lack of access to comprehensive healthcare which has been one of their most prominent demands (Berkins and Fernández 2005; Berkins 2007; Red LACTRANS 2014; ATTTA and Fundación Huesped 2014). However, some changes have occurred in recent years. Following the civil marriage reform of 2010,[1] a progressive Gender Identity Law was voted in 2012.[2] This law guarantees universal and free-of-charge access to hormonal and surgical procedures as well as legal recognition of self-perceived gender identity for transgender individuals.

These changes in legislation and policies opened up a new scenario for the relationship between transgender individuals and healthcare services, historically marked by exclusion and symbolic persecution towards the trans community. Thereby, this chapter analyses the experience of waiting in healthcare services taking into account this new context. We will argue that the biomedical and 'cis'[3] shaped gaze present in healthcare services—as part of a context of historical, social and institutional exclusion—contributes to the production of a different temporality in the transgender waiting experience.

---

[1]The law N° 26.618, recognizes the right to marry and adopt for gay and lesbian couples.
[2]The law N° 26.743 was approved in 2012. Within the Ministry of Health, this law has been implemented since 2015.
[3]The prefix 'cis' (as in cisgender) refers to the power structure, which considers 'normal' and hierarchizes individuals whose inner experience of gender corresponds to the assigned sex at birth. 'Cis' functions as the antonymic pair of 'trans'.

Methodology

This exploratory investigation has been carried out through a qualitative approach (Glaser and Strauss 1967). We conducted in-depth interviews with transgender people and medical practitioners (endocrinologists), since the interview is a privileged resource to collect the actor's perspective (Alonso 1998). This technique is adequate to identify practices and opinions of medical practitioners about transgender patients, and it is useful to reconstruct 'waiting' experiences of transgender individuals. We interviewed female and male patients who had undergone hormonal and/or surgical treatments in medical institutions. Recruitment was facilitated through institutions and groups of activism that bring together transgender people, as well as the use of 'snowball' sampling.

Regarding the selection of the two medical practitioners (endocrinologists), this selection was based on their experience and frequent prescription of hormonal treatments to transgender patients in public hospitals. Throughout the study, ethical and methodological precautions were taken into account to protect the identity of each person interviewed. We also conducted observations at public healthcare services[4] in order to gather more information about the scenarios in which transgender 'waiting experience' in public medical institutions takes place.

One of the main characteristics of public hospitals in Argentina is that they are free-of-charge and offer universal assistance to those who lack private or union-based health insurance. In each of the two hospitals we observed, the services that assist transgender population are composed of an interdisciplinary team of psychologists, clinicians and endocrinologists. The observations took place at two of the hospitals' waiting rooms that deliver care to this population in Buenos Aires.

For the analysis of the interviews, we progressively developed some categories, which were later integrated into the theoretical scheme that allowed describing their relationships and explaining the phenomenon

---

[4]This project is part of a larger study directed by Mario Pecheny and funded by the University of Buenos Aires and CONICET, in which various situations of waiting are examined in the fields of health, money and love.

studied in a coherent way (Braun and Clarke 2006). We particularly focused on the transgender interviewees' narrative descriptions of the 'scenes', following Vera Paiva's (2012) proposal. As she points out, this technique is meant to set in motion a process of defamiliarization directed at certain cultural scenes structured through scripts internalized in each of us through primary socialization. This can be useful to comprehend different dimensions, dynamics and significant elements of waiting experiences that might even provide insights into larger social and political analysis.

The Gender Identity Act in Argentina's Healthcare System

In Argentina, transgender community activism has been progressively more visible since the restoration of democracy in 1983. In the early 1990s, these organizations began to wage a fight to be formally recognized by the State. In 1996, the demand for recognition of gender identity as a right, understood as a separate category from sexual orientation, was formally introduced during the constituent assembly of Buenos Aires (Berkins 2003). This started a process of deliberation within the transgender leaders and organizations in order to generate a legal reform, which would eventually lead to the enactment of the gender identity law. This law bases its understanding of gender identity on the Yogyakarta Principles (2007). According to these principles, gender identity refers 'to each person's deeply felt internal and individual experience of gender, which may or may not correspond with the sex assigned at birth, including the personal sense of the body (which may involve, if freely chosen, modification of bodily appearance or function by medical, surgical or other means) and other expressions of gender, including dress, speech and mannerisms'.

The Gender Identity Law approved in 2012 does not presume the notion of any pathology in transgender bodies or any mental disorder as a requirement to the legal recognition of gender identity or to universal access to medical procedures.⁵ Furthermore, people can access free hormonal treatment and/or surgical treatments without any judicial authorization requirement (Farji Neer 2014; Ortega 2015). This should signify an improvement in the bureaucratic procedures, as well as, in the waiting time required for a transgender individual to access hormonal and surgical treatments according to his or her self-perceived gender identity.

Despite the legislation's approval in 2012, its implementation and the recognition of its concrete effects on people's lives do not constitute a rapid and linear process. Indeed, its Article 11⁶—regarding the right to comprehensive health care—was not regulated by the Ministry of Health until May 2015.

In this context, our chapter presents the experiences of transgender individuals while waiting for healthcare attention as related to accessibility to treatments. Accessibility, understood as the link between subjects and public services, is defined as 'the conditions and guidelines of a service, and the conditions and representations of the subjects that reveal itself through the use of the service' (Stolkiner and Barcala 2000).

Hormonal treatments are not the only ones allowed by Argentinean identity law, since both male and female transgender individuals might require surgical procedures. However, we found in the initial phase of our investigation that all of our interviewees had received hormonal treatments, both with and without medical supervision. Previous studies have highlighted the dangers that transgender people face when bodily modifications such as silicone injections, prosthetic implants and HORMONE use are performed without professional supervision (Berkins 2007). Due to exclusion from hospitals and the lack of access to drugs and professional advice, transgender people carried out these 'homemade' treatments and practices in a non-sterilized environment or without medical supervision. Moreover, interviewees prominently cited self-medication

---

⁵For example, in several countries 'gender dysphoria' and 'gender identity disorder' are diagnoses necessary to change a person's official identity gender. This is not the case in Argentina.
⁶The legal order listed the interventions that can 'help match the body to the self-perceived gender identity' (Decree 903/2015, Annex I point 1) and provided rules and guidelines around their inclusion in Argentina's mandatory public healthcare plan, which regulates services covered by public health, union-based and private sectors.

and the discontinuation of hormonal treatments as consequences of inadequate access to health care.

We have chosen to focus on the public health sector since it is typically used by people who have no other form of medical insurance (union-based or private). In fact, the public sector is the only access to health care in Argentina for poor and unemployed citizens. In this scenario, waiting becomes crucial for accessing medical services. Thus, we locate the analysis of transgender patients' waiting experiences while seeking hormonal therapy within public hospitals of Buenos Aires.

The Experiences of Waiting

Considering Foucault's (1980) relational conception of power, we understand waiting, and specifically the waiting/making wait dynamic, as a power phenomenon. Therefore, studying scenes of waiting and making others wait allows us to understand power's effects and its circulation, including the subjectification involved in waiting. We privileged the use of scenes as a methodology that, through a detailed reconstruction of the scenery of waiting, provides information about significant elements on waiting periods. Accordingly, the dimensions we take into account refer to: time of waiting (including temporality), forms of interaction, elements that constitute the scene, body and embodiment of waiting, among others.

By focusing on subjective dimensions, we seek to analyse the experience of waiting. As Joan Scott explains (2001), experiences can be used to reconstruct and give meaning to socio-cultural processes. The waiting experience does not only indicate the chronological time that stands between a subject and a service. Since experiences have meaning only in relation to the broader context in which they take place, analysing transgender patients' waiting experiences also serves as a rich place for us to examine power relations and their role in the construction of gendered temporalities.

The act of waiting to have access to a service can be conceptualized as a form of political subordination. Javier Auyero (2013) has studied the experiences of people who seek social assistance from the local government of Buenos Aires. These people need to complete several steps and undergo long waits in order to obtain an appointment with the public officer who might or might not answer a claim. The arbitrary and uncertain character of these waiting periods, and the power of some to 'solve' the waiting, is what allows us to describe this power as a mechanism of political domination. Auyero affirms that the Ministry of Social Development doesn't treat economically poor people as citizens, but rather as 'patients of the State' with the implicit demand of submission (2013, 39). He explains how state employees devalue the time of lower-class citizens, which can be read as a demeaning of people themselves.

Using Auyero's work, we define *waiting periods* as 'temporal processes through which political subordination is reproduced' (2013, 16). In this way, an individual waiting experience is indicative of power dynamics that exist between the subject and the state. The waiting experiences of patients show how the State and public health institutions recreate and reproduce the conditions of vulnerability to which individuals and groups are already exposed. This dimension of vulnerability can be defined as 'programmatic' (Ayres et al. 2012; Pecheny 2013), because it is produced by the actions and/or lack of actions from State institutions.

The field of social studies of emotions has also considered waiting experiences as the object of analysis. According to Adrián Scribano (2010), capitalist domination produces mechanisms that prevent conflict and make this domination emotionally bearable. One such mechanism is the act of waiting because it requires people to develop patience. It is the only possible strategy for people from disenfranchised classes, for whom acting as a 'patient citizen' is a sort of medicalization of political life (Scribano 2010).

A concept of 'generic waiting' (Day 2016) entails the control of anxiety, the completion of bureaucratic processes, and the 'abandoning of the body' (Scribano 2010). But not all bodies can be 'abandoned' to the same degree. Transgender bodies, which exist in a context of gender stigmatization, cannot escape the judgemental gaze of the cis-normative, heteronormative society. This is an example of how transgender bodies are forced to face and overcome specific obstacles in situations of waiting. The hierarchy that exists between cis (i.e. non-transgender) and transgender bodies informs a person's subjective experience of waiting.

Despite these predominant dynamics, waiting also entails different dimension of social life where agency is highlighted (Hage 2009; Dwyer 2009; Minnegal 2009). For instance, our interviewees used different strategies of self-medicalization, decided whether to wait or not to receive health care and where to do so, among other examples. In that way, we not only emphasize that transgender people are shaped by those who make them wait, but also they shape that waiting, too. Moreover, patience associated with waiting can also be seen as a process leading to political organization. This is what Appadurai calls a 'politics of patience', meaning 'the capacity of people to transform their needs into a collective claim, and to patiently work to organize themselves into a long-term project that endeavours to confront and solve this claim' (Appadurai in Bandak and Janeja 2018, 8). This process can be linked with the political organization of transgender people in Argentina, which, as stated above, lead to passing the gender identity law.

Waiting Scenes: Public Health Services

On a cold winter morning, we visited one of the busiest hospitals in Buenos Aires. The objective was to identify and know the location, and to interview some patients and doctors in the transgender health sector. After explaining our research and our interest in interviewing somebody from the medical team to the receptionist, she told us that we would have to wait to talk to the chief of medical service. After a few moments, a doctor came out and told us that he was not sure if he would have time to answer our questions. He asked us to wait a little longer or he could meet us another day. After waiting for an hour, he came out of his office and told us it would be impossible for him to meet with us. Thus, we arranged an appointment for another day.

In the meantime, while we were waiting, we were able to make several observations. It was evident that the available seats were not enough for everyone to sit down in the waiting room. Also, there was a lot of traffic since people were coming from all sides of the hospital (the hallways, the stairs and the elevators). A few people were standing hesitantly, asking for directions or reading the signs on the wall trying to find a service or a laboratory. The endocrinology service, which takes care of transgender patients, is not delimited to a specific area. It shares a common space with other large sectors such as Urology, Gynaecology and Cardiology. We realized that this kind of space distribution—the high level of exhibition in this common area—can be particularly uncomfortable for transgender patients and may have a negative impact on their access to professional care (Fig. 4.1).

We saw no signs indicating the endocrinology service. There was no available information about where and how the service functioned. This lack of information contributes to the hierarchical gap between doctors and the population they assist. However, we did find examples of bureaucratized relationships between doctors and patients, as the following sign shows (Fig. 4.2).

Waiting periods generate feelings of vulnerability and discomfort. A transgender man we interviewed gave a detailed description of his first visit to the gynaecologist:

> To be honest, the experience (of waiting) was quite tense because there was the whole waiting room and there was only one main front desk

Fig. 4.1 Urology Division

Fig. 4.2 Patients with appointments to be hospitalized, form a line here

and it was filled with women who stared at me and didn't understand what I was doing there. Even though, technically, men can also go to the gynecologist even if they aren't trans, like when they have problems with their mammary glands or something like that. But since that isn't common knowledge, all of the gazes were fixed on me to try and figure out what I was doing there… like if I was waiting for someone who was in a consultation, or if I was waiting for someone to arrive… (Pedro, age 21)

He also described the scene at the moment when the secretary called his name to register him as a patient of a gynaecologist:

So that was a whole other issue, because she was telling me 'but give me the ID of the person who is coming to the appointment' and I said: 'Well … I'm the person who's coming to see the doctor'. I explained to her I was transgender… but like she didn't get it. She went inside, asked someone, and when she came back out she told me, 'Wait, so are you the one coming for an appointment? I don't understand.'

Our interviews with transgender patients shed insight on how the conditions of the hospital infrastructure transformed waiting periods into instances of discipline, discomfort and programmatic vulnerability. This may explain why many of our interviewees said they preferred to go to medical consultations alone, without any family or friends. Some interviewees, knowing in advance this situation, prevented others from the effects of courtesy stigma (Goffman 1963/2012). Even though this scene describes one of the despicable situations that transgender people have to undergo to access healthcare assistance; that kind of anticipation shows how transgender people manage and resist the stigma that certain waiting experiences creates, as the following quote shows:

> It's a situation… I knew it was going to be a very uncomfortable situation and I didn't want to be with another person who would also get uncomfortable, I preferred to be alone… (Juan, age 25)

We observed similar waiting scenes at another public hospital in Buenos Aires. The waiting room was crowded, there was very little information on how to navigate the health services and there were long lines to schedule an appointment—the earliest of which was not before a month. We also noticed irregularities in the distribution of appointments that responded to discretionary management. This involved certain 'cronyisms' that existed between some patients and people working behind the desk (Fig. 4.3).

A transgender woman told us about her waiting experience in a hospital, some years before the Gender Identity Law was passed. The interviewee's narrative highlights how the discretionary mechanism that permits the unequal distribution of appointments is strongly informed by prejudice and discrimination:

> Seven years ago, I came to this hospital and I was not cared for in the way that they care for people right now…I suffered from this discrimination and the discrimination led me to wait longer because of what the doctors and nurses thought at that time… well, it was what it was… my presence was ignored… it was like… when they decided that it was my turn, it would be my turn (…) I got there, I signed up, I think I was the fifth or sixth name on the list, and I was called like fifteenth or something. Every

Fig. 4.3 Waiting room

> time I went to ask why they hadn't called me, they avoided my question telling me that there were already appointments scheduled before me… even though I saw that people got there, signed up, and were called up before me… (Romina, age 36)

Although this situation is a clear example of how frustrating waiting for health care is for transgender people, this scene shows that waiting can also vary (Day 2016). Even when the result might be the same as those who are waiting in a 'passive' way—that is continue waiting for longer periods of time—this active attitude shows that despite restricted and unidirectional power, there are still certain ways of resisting and confronting that power.

In addition to the long wait itself, to the hours spent, these supplementary conditions of insecurity and feelings of injustice affect the quality, adherence and the continuity of medical care. Furthermore, these conditions negatively impact the feeling associated with this waiting experience:

> Waiting too much for an appointment, you're just like 'ok, it feels like I am spending my entire life trying to get a stupid appointment.' And a lot of that has to do with why I ended up giving up on getting treatment… I told myself, OK, I'll continue it sometime in the future, when things will work a little better, or there will be other places set up specifically for this. (María, age 23)

Other interviewees reported similar feelings of frustration when they tried to access medical care. Interactions with administrative staff reproduce exclusions that affect the transgender population in other areas of society, such as their exclusion from housing and from the education system:

> Awful, awful… I felt like… Really that day I felt like I didn't have a place in the world, like… that day I asked myself 'what am I doing here?' If I have to go through all of this… not only here… we carry a heavy social burden by being excluded from all places. Even though I did not get kicked out of my house like many, but I did have to go through these exclusions in schools, in hospitals… all the systems which I have and will have to go through… well except for police stations and prison cells… there we are always included… (Romina, age 36)

Health institutions impose disciplinary practices that shape the experience of waiting in a way that hinders accessibility to services, practitioners, drugs and treatments. In the case of transgender people, these practices manifest a biomedical gaze that structurally excludes non-cisgender bodies. Our interviews described these waiting periods as paradigmatic scenes that are dreaded and feared by the transgender community at large. A specifically traumatizing circumstance of these waiting experiences is that transgender individuals have been historically called in a loud voice by their registered name, which may not match the person's gender identity and expression. Being misgendered is perceived as a form of violence because it can be deeply humiliating for transgender people. Although the Gender Identity Law has recently recognized the right for people to match their name and gender to their self-perceived identity, the fear and indignation aroused in these scenes in hospitals—and other institutions—(Berkins 2007) remain alive in the emotional memory and current experiences of the transgender community.

The Transgender Population Seen Through the Biomedical Gaze

An increasingly body of literature on transgender populations characterizes the different barriers to access high-quality health care. These barriers include ways of discrimination related to real or perceived stigma of gender nonconformity (Bockting et al. 2004; Cobos and Jones 2009), structural and financial barriers and socioeconomics barriers (transportation, housing), among others, but the main barrier is the lack of access to care which creates health disparities (Safer et al. 2016). The historical marginalization and transphobia contribute to perpetuate this disparity (Roberts and Fantz 2014). This lack of health insurance and living in poverty makes transgender people more likely to delay care (Cruz 2014). These situations have created healthcare inequality for the transgender population (Edmiston 2018). Despite the prominent literature regarding transgender population in Argentina, there is still little research on the experiences of transgender people while seeking for healthcare attention. Regarding this issue, the interviews addressed another factor that may bar transgender people from accessing care and treatments: the tendency of doctors to make decisions on behalf of them. As many interviewees explained, some doctors were reluctant to see them, either because they lacked knowledge of transgender health, or because they had prejudice against them. Ignorance, misinformation and negative prejudice held by health professionals (medical doctors, surgeons, nurses, psychologists, social workers, etc.) are informed by the hegemonic biomedical model. This dynamic is reaffirmed through the submissive attitude that patients must show as an implicit requirement of the patient-doctor relationship. Consequently, patients are largely excluded from knowledge about their health and doctors tend to pathologize problems through their medicalization (Menéndez 2003).

Historically, medicine has been a key field in the enforcement of the gender binary. For example, doctors often perform gender reassignment surgery on intersex individuals to 'fix' people with ambiguous genders (Cabral and Benzur 2005). The pathologizing of transgender identities is still widely expressed in manuals and taxonomies used globally, through the enumeration of various psychiatric disorders (transsexualism, gender identity disorder, transvestic fetishism, gender dysphoria) (Missé and Coll-Planas 2010; Farji Neer 2014; Ortega 2015). This hegemonic model conceptualizes the relationship between sex and gender as coherent, fixed and strictly biological. Social, psychological and cultural processes that shape one's identity are regarded as anecdotal, and without real medical significance.

Biological determinism and binarism shaped the discourse of one of the first doctors in charge of treating the transgender population in a major general hospital of Buenos Aires. During the interview, she showed us a text message that one of her patients had sent to her phone and said:

> Look, this transgender boy is one of my patients... he's a boy, but he still has these girl traits, as you can see from the little hearts that he's puts in his messages. (Endocrinologist, age 42)

In the same hospital, a transgender patient described to us what happened to him during a visit with a doctor. The doctor told him:

> Your levels of testosterone are higher than they should be. A woman does not have this amount of testosterone. Have you already taken some? I told him no, I'd never taken hormones, and he said: 'Good, well that is the problem then. I'm going to give you some estrogen pills that way your testosterone will go down and all those ideas of wanting to be a man will go away.' (José, age 23)

According to the interviews, the initial waiting periods, loaded with anticipated fear of being discriminated, as well as the negative attitudes of doctors thereafter, may act as barriers to comprehensive care. The insecurity and anguish that many transgender individuals experience, both in anticipation to and during medical care, can be drastically reduced once a relationship of trust and respect is established between the patient and the professional.

An interviewee commented that during his second visit to his gynaecologist:

> ... I was going to the appointment with a completely different mindset because even though I had to wait, I knew that I was going to go in and everything was going to be good. (Pedro, age 21)

Waiting can be thought as an expression of agency; for example, when we choose to wait because it is convenient for us (Minnegal 2009). Taking this into account, even though a waiting period is implicit in the scene described above, this waiting is signified by the interviewee as one that is worth it to be waited. In fact, he shows certain wilfulness to wait. The main reason of choosing waiting, in this case, is related to the trust in the professional. For many patients, the waiting experience brings certain hope of finding someone that can solve their problems, or represent someone that 'might also be thinking about you in between visits' (Day 2016, 170). Hope, or an imagined future, can certainly affect the ways people wait re-shaping the waiting experience (Bandak and Janeja 2018; Brun 2015). In these cases, as this verbatim shows, waiting periods are signified positively.

Related to this, one of the doctors told us that the long waiting time to receive treatment was explained by the comprehensive treatment that this hospital provides. Apparently, there seem to be certain expectations on behalf of patients to be taken care of. According to him, the oversight is much more interdisciplinary in this hospital than in other institutions. He said that:

> People giving up before showing up? I have no idea. In general, the ones who sign up come. It is very rare... Some people go to other hospitals, and they see that the waiting list is shorter, but they still come here, because the care we give here is much more comprehensive and we have an important team (...) The care that we provide here, no one else provides it. (Endocrinologist, age 50)

Through interviews with the doctors, we also noticed certain conceptions about transgender people as 'less patient', as one of the doctors commented:

> (…) they are a population with a *particular patience*… maybe it is the fact that then know that worst comes to worst, they can go and buy hormones in the pharmacy and they can do it themselves. They don't have the same necessities as a person who is sick and needs a diagnosis.

And regarding the continuity of the treatment, the endocrinologist added:

> I don't know, they're not very adherent (…) adherent in relation to check-ups and coming to get treated.

As this quote points out, there is certain (cis)temporality which obliterates the effects of stigmatization. This cisgendered temporality conceives transgender people and its 'particular patience' as impatient and thus, as 'bad patients'. However, this conception does not contemplate other factors that might have some influence in this phenomenon, such as economic factors. As many literatures on transgender healthcare points out, the lack of health insurance and living in poverty conditions—which are this population's main conditions—tends to delay care (Cruz 2014).

The comparison between doctor's opinion and the Royal Spanish Academy dictionary's definition of the word *paciente* (patient) is interesting:

1. That has patience.
2. It is used to refer to a subject that receives the action of the agent.
3. It is a person that receives the action of the verb.
4. A person that suffers physically, especially one under medical care.
5. A person who is or will be medically recognized.

The notion of patience is inherent in the registered definition of the Spanish word *paciente* (patient). To be patient—or having patience—as Gasparini (1995) noticed, implies giving space to other conceptions of temporality as slowness. This contrasts with cultural hegemonic models, which value positively efficiency and rapidity. According to these terms, 'the attitude of patience expresses the full acceptance of the other's time, which cannot be reduced to our time' (Gasparini 1995, 42). Patience, as a prerequisite to access healthcare services, informs the expectations that professionals have of patients. This can explain the recognition of transgender people as not very committed to hormone treatments, or 'not adherent'. In addition, it is noteworthy that the contingent nature of the diagnosis is mirrored by the construction of the 'other' as 'irresponsible', 'selfish', for their lack of medical consistency. Thus, by erasing the effects of stigma and discrimination from the experience of waiting—a cis conception of time—the conversation around waiting places implicit responsibility on the people who wait. Patients are the ones that do or do not possess the quality of patience. This mechanism hides social and institutional power structures that articulate the waiting periods.

The impatient trans person created in this structure is also constructed as a patronized (infantilized) category, like children, since children are a typical model for the impatient. This is apparent in the fact that transgender adults were commonly referred to as 'boys' and 'girls' by health professionals we interviewed. From a judicial perspective, 'boys' and 'girls' are minors that require guardians to make decisions on their behalf until they are old enough to be legally autonomous (Varela et al. 2005). Referring to transgender people as 'boys' and 'girls', undermine an understanding of transgender people as legal possessors of right. Furthermore, this infantilized definition is informed and reproduced through paternalistic practices that the biomedical model imposes on transgender bodies. This model constructs the transgender person as an 'other' for doctors and, this othering of the trans patients, allows them to be defined through this series of stigmatized particularities.

Concluding Remarks

Throughout this chapter, we analysed waiting experiences of transgender patients accessing public healthcare services in the context of the recently passed Gender Identity Law. The interviews and observations helped us to conceptualize trans waiting periods as a scene, usually marked by anguish and fear which may in turn act as institutionalized barriers that uphold the exclusion of the transgender community. We also explored how patience, which is tacitly or explicitly mandated, operates as a subtle mechanism to discipline and control bodies. By identifying dimensions and significant elements of the dynamic of 'waiting / making wait' that take place at public healthcare system, we arrive to preliminary conclusions of waiting experiences among the transgender population.

Waiting experiences of transgender people have several unique qualities. While waiting periods may usually be seen as a trigger for boredom and anxiety, the interviewees described feelings of fear, unfairness, humiliation and anguish. Those emotional descriptions constitute an experience of a different temporality in transgender people involved in this waiting experience. The singularities of this experience can be attributed to historic stigmatization and discrimination suffered by this community. Taking this background into account, we can unveil a deeper meaning for the feelings that arise among transgender people while waiting at a hospital. The waiting room in a hospital has been (and still is) a scene where those who step aside from a cisgendered system can be severely judged by the audience. Here, the definition of transgender bodies in the biomedical sphere plays an important role. In order to analyse the power of 'making waiting' in Foucault's perspective, it is also necessary to consider a wider political and historical structure. Hence, the history of pathologizing transgender people functions as a scenography for the conception of these subjects as impatient and 'bad patients'. This medical gaze, materialized in the discourses of health professionals, now also uses a "paternal language" that continues othering and stigmatizing transgender subjects.

Besides the paternalistic practices and the misconception about transgender people as 'childish' held by doctors, during the interviews we also recognized a certain negotiation enacted by the patients. This was evident, for example, when one of the doctors referred to their practices of self-medication. Furthermore, some transgender interviewees hinted that they decided whether to continue or not with the treatment according to their perception of the waiting experience as one worth the wait or not. These situations show power, not only in one direction, but also as a relational dimension. Moreover, they are a manifestation of how transgender people reconfigure power relations while seeking to access to health care.

The Gender Identity Law provides a legal framework to deconstruct the pathologizing of transgender identities. Furthermore, it provides an ambitious blueprint to revolutionize their access to comprehensive health care. Although the law and its implementation represent a substantial step forward, the agency and mobility of transgender people in the healthcare system are still strongly limited. This becomes evident by considering waiting experiences of transgender people in a wider framework. Reconceiving temporality by criticizing the cis-shaped view of the world may be a part of a larger effort to continue deconstructing the gender binarism used to conceptualize transgender bodies in the dominant biomedical discourse.

References

Alonso, L. E. 1998. *La Mirada Cualitativa en Sociología*. Madrid: Fundamentos.
Asociación de Travestis, Transexuales y Transgéneros de Argentina (ATTTA) y Fundación Huésped. 2014. *Ley de Identidad de Género y acceso al cuidado de la salud de las personas trans en Argentina*. Buenos Aires: Fundación Huésped.
Aday, L. A., and R. Andersen. 1974. "A Framework for the Study of Access to Medical Care." *Health Services Research* 9 (3): 208–220.
Ayres, J. R., V. Paiva, and I. J. Franca. 2012. "Conceitos e Praticas de Prevenção: Da Historia Natural da Doença ao Quadro da Vulnerabilidade e Direitos Humanos." In *Vulnerabilidade e direitos humanos. Prevenção e promoção da saúde: Vol. I. Da doença a cidadania*, Chap. 1. San Pablo, SP: Jurúa.
Auyero, J. 2013. *Los Pacientes del Estado*. Buenos Aires: Eudeba.
Bandak, A., and Manpreet K. Janeja. 2018. "Introduction: Worth the Wait." In *Ethnographies of Waiting: Doubt, Hope and Uncertainty*, edited by M. K. Janeja and A. Bandak, 1–39. London, Oxford, and New York: Bloomsbury Academic.

Berkins, L. 2003. "Un Itinerario Político del Travestismo." In *Sexualidades migrantes. Género y transgénero*. Buenos Aires: Scarlett Press.

———. 2007. *Cumbia, Copeteo y Lágrimas. Informe Nacional sobre la Situación de las Travestis, Transexuales y Transgéneros*. Buenos Aires: ALITT.

Berkins, L., and J. Fernández. 2005. *La Gesta del Nombre Propio. Informe sobre la situación de la comunidad travesti en la Argentina*. Buenos Aires: Asociación Madres de Plaza de Mayo.

Bockting, W., B. Robinson, A. Benner, and K. Scheltema. 2004. "Patient Satisfaction with Transgender Health Services." *Journal of Sex & Marital Therapy* 30 (4): 277–294. https://www.ncbi.nlm.nih.gov/pubmed/15205065.

Braun, V., and V. Clarke. 2006. "Using Thematic Analysis in Psychology." *Qualitative Research in Psychology* 3 (2): 77–101. https://doi.org/10.1191/1478088706qp063oa.

Brun, C. (2015) "Active Waiting and Changing Hopes: Toward a Time Perspective on Protracted Displacement." *Social Analysis* 59 (1): 19–37.

Cabral, M., and G. Benzur. 2005. "Cuando Digo Intersex: Un Diálogo Introductorio a la Intersexualidad." *Cadernos Pagu* 24: 283–304.

Cobos, D. G., and J. Jones. 2009. "Moving Forward: Transgender Persons as Change Agents in Health Care Access and Human Rights." *Journal of the Association of Nurses in AIDS Care* 20 (5): 341–347. Available in: http://refhub.elsevier.com/S0277-9536(14)00211-1/sref10.

Cruz, T. M. 2014. "Assessing Access to Care for Transgender and Gender Nonconforming People: A Consideration of Diversity in Combating Discrimination." *Social Science & Medicine* 110: 65–73. https://doi.org/10.1016/j.socscimed.2014.03.032.

Day, S. 2016. "Waiting and the Architecture of Care." In *Living and Dying in the Contemporary World: A Compendium*, edited by V. Das and C. Han. Oakland, CA: University of California Press.

Dwyer, P. D. 2009. "Worlds of Waiting." In *Waiting*, edited by G. Hage, Chap. 1. Carlton, VIC: Melbourne University Press.

Edmiston, E. K. 2018. "Peer Advocacy for Transgender Healthcare Access in the Southeastern United States: The Trans Buddy Program." In *Healthcare in Motion: Immobilities in Health Service Delivery and Access*, edited by C. Vindrola-Padros, G. A. Johnson, and A. E. Pfister. New York: Berghahn Books.

Farji Neer, A. 2014. "Las Tecnologías del Cuerpo en el Debate Público. Análisis del Febate Parlamentario de la Ley de Identidad de Género Argentina." *Sexualidad, salud y sociedad* 16: 50–72.

Foucault, M. 1980. *Microfísica del Poder*. Madrid: Gedisa.

Gasparini, G. 1995. "On Waiting." *Time & Society* 4 (29): 29–45. doi.org/https://doi.org/10.1177/0961463x95004001002.
Glaser, B., and A. Strauss. 1967. *The Discovery of Grounded Theory: Strategies for Qualitative Research.* Chicago: Aldine.
Hage, G. 2009. "Introduction." In *Waiting*. Carlton, VIC: Melbourne University Press.
Ley 26.743 de Identidad de Género. 2012. Viewed 20 October 2019. Available in: http://www.infoleg.gov.ar/infolegInternet/anexos/195000-199999/197860/norma.htm.
Ley 26.618 de Matrimonio Civil. 2010. Viewed 20 October 2019. Available in: http://servicios.infoleg.gob.ar/infolegInternet/anexos/165000-169999/169608/norma.htm.
Goffman, E. [1963] 2012. *Estigma: La Identidad Deteriorada.* Buenos Aires: Amorrortu.
Menéndez, E. 2003. "Modelos de Atención de los Padecimientos: De Exclusiones Teóricas y Articulaciones Prácticas." *Ciencia & Saúde Colectiva* 8 (1): 185–207.
Minnegal, M. 2009. "The Time Is Right: Waiting, Reciprocity and Sociality." In *Waiting*, edited by G. Hage, Chap. 7. Carlton, VIC: Melbourne University Press.
Missé, M., and G. Coll-Planas. 2010. *El Género Desordenado. Críticas en Torno a la Patologización de la Transexualidad.* Madrid: Egales.
Ortega, J. 2015. "Sobre la Exigibilidadas del Derecho a la Salud en Personas Trans: De Conquistas y Deudas aún Pendientes." Paper presented at the VII Congreso Internacional de Investigación y Práctica Profesional en Psicología, UBA. Buenos Aires, November 25–28.
Paiva, V. 2012. "Cenas da Vida Cotidiana: Metodologia para Compreender e Reduzir a Vulnerabilidade na Perspectiva dos Direitos Humanos." In *Vulnerabilidade e direitos humanos. Prevenção e promoção da saúde: Vol. I. Da doença a cidadania*, Chap. 8. San Pablo, SP: Jurúa.
Pecheny, M. 2013. "Desigualdades Estructurales, Salud de Jóvenes LGBT y Lagunas de Conocimientos: ¿Qué Sabemos y Qué Preguntamos?" *Temas em Psicología* 21 (3): 961–972.
Pecheny, M. 2017. "Introducción." In *Esperar y Hacer Esperar: Escenas y Experiencias en Salud, Dinero y Amor*, comp. by Mario Martín Pecheny and Mariana Palumbo. Ciudad Autónoma de Buenos Aires: Mario Martín Pecheny.
Red LACTRANS. 2014. *Informe sobre el acceso a los derechos económicos, sociales y culturales de la población trans en Latinoamérica y el Caribe.* Viewed

20 October 2019. Available in: http://redlactrans.org.ar/site/wp-content/uploads/2015/03/Informe%20DESC%20trans.pdf.

Roberts, T. K., and C. R. Fantz. 2014. "Barriers to Quality Health Care for the Transgender Population." *Clinical Biochemistry* 47 (10–11): 983–987. https://doi.org/10.1016/j.clinbiochem.2014.02.009.

Safer, J. D., E. Coleman, J. Feldman, R. Garofalo, W. Hembree, A. Radix, and J. Sevelius. 2016. "Barriers to Healthcare for Transgender Individuals." *Current Opinions in Endocrinology Diabetes and Obesity* 23 (2): 168–171.

Scott, J. 2001. "Experiencia." *La ventana* 2 (13): 42–74.

Scribano, A. 2010. "¡Primero hay que saber sufrir! Hacia una sociología de la 'espera como mecanismo de soportabilidad social'." In *Sensibilidades en juego: miradas múltiples desde los estudios sociales de los cuerpos y las emociones*, Chap. 7, 169–192. Córdoba: Estudios Sociológicos Editora.

Siciliani, L., V. Moran, and M. Borowitz. 2014. "Measuring and Comparing Health Care Waiting Times in OECD countries." *Health Policy* 118 (3): 292–303. doi:https://doi.org/10.1016/j.healthpol.2014.08.011.

Stolkiner, A., and A. Barcala. 2000. *Reforma del Sector Salud y utilización de servicios de salud en familias NBI: estudio de caso. La Salud en Crisis - Un análisis desde la perspectiva de las Ciencias Sociales*. Buenos Aires: Dunken.

Varela, O., O. Sarmiento, S. M. Puhl, and M. A. Izcurdia. 2005. *Psicología Jurídica*. Buenos Aires: JCE Ediciones.

Yogyakarta Principles. 2007. *Yogyakarta Principles on the Application of International Human Rights Law in relation to Sexual Orientation and Gender Identity*. Buenos Aires: Editorial Jusbaries.

5

Living in 'Limbo': Immobility and Uncertainty in Childhood Cancer Medical Care in Argentina

Eugenia Brage

Introduction

In Argentina, cancer is the leading cause of death by illness in children from five to fourteen years, and the second in children from zero to four years (DEIS 2010). Around 1290 new cases are diagnosed per year. According to the National Cancer Institute, approximately 400 children die annually of cancer, recognizing as main causes the 'late diagnosis, the difficulty in the referral in time and form, the complications in the treatment and in some cases, the lack in the integral care of the patient' (INC, n.d.). Although the percentage of cure is similar than in developed countries—65%—(Abriata and Moreno 2010), the estimated percentage of survival is significantly lower (Moreno et al. 2015), with substantial differences depending on the region of residence (Felice et al. 2013; Aurelie Pujol et al. 2014). In this way, the prognosis of a child diagnosed

E. Brage (✉)
Center for Metropolitan Studies (CEM),
University of São Paulo, São Paulo, Brazil

with cancer varies according to the geographical region of the province of residence (ROHA 2012).

Approximately half of the children affected by cancer must move for indeterminate periods to receive diagnoses or medical attention in specialized hospitals (ROHA 2012). These transfers usually occur within the framework of a medical referral[1] from hospitals of medium and low complexity located in different provinces, towards highly complex hospitals located, for the most part, in Buenos Aires (Abriata and Moreno 2010). Despite the existence of these networks, in the initial stages late referrals and diagnostic delays have been identified leading to a 'loss of therapeutic opportunities' (Abriata and Moreno 2010, 45). This gives rise to transfers outside the institutionalized referral networks, which means that a large number of children and their primary caregivers move on their own, outside these institutional networks, searching for solutions they do not find at their local contexts. The lack of resources for diagnoses and treatments for the therapeutic approach of this disease are the main factors that lead to a medical referral. Distrust in clinical management, lack of responses in places of origin and hope to find 'better care' represent the main factors that motivate people to travel by their own means (Vindrola-Padros and Brage 2017).

Taking this into consideration, migration is central in the therapeutic itineraries of children with cancer (Brage 2018). However, despite the decisive importance of these mobility phenomena in the course of disease, it is often considered a 'secondary aspect' in a biomedical context where there is a prevalence of biological approaches that consider disease in its material aspect. The aim of this paper is to show the centrality of immobility in cancer medical care arguing that migration and disease are not isolated but, instead, they constitute a unit of experience (Turner and

---

[1] Medical referral is the process of transferring a patient from one service to another or from a hospital or health centre to another of higher complexity. In Argentina, the health system has autonomy in the different provinces and is organized by levels of care according to the degree of complexity of the disease. The first and second levels are characterized by being decentralized in provinces regulated by their respective jurisdiction. The third level of care, which includes institutions specializing in highly complex diseases, is mostly centralized in Buenos Aires, despite the fact that decentralization has been promoted in the last decade based on the construction of hospitals in the different provinces, the training and formation of human resources and the implementation of communication networks.

Bruner 1986), framed in contexts of structural violence (Farmer 2007). The transformations produced by the geographic displacement and the reconfiguration and adaptation to a new migratory context are combined with hospitalization as a daily life and the impossibility of returning to the place of origin, which configures a prolonged liminal state. In this way, understanding migration from immobility perspectives and considering uncertainty as a central aspect in chronic diseases studies, this paper shows that childhood cancer medical care in Argentina give rise to a liminal phase which I have called 'limbo', linked to structural factors that prolong and increase suffering.

Ethnographic fieldwork was carried out from May 2013 to December 2015 in three main areas: a public paediatric hospital located in Buenos Aires, a non-governmental organization (NGO) created to support the low-income families that go through this disease and, finally, a hotel where some families are hosted during the period of medical treatment in Buenos Aires. The research project was evaluated and approved by the ethics committee of the hospital, granting me authorization to develop my fieldwork activities. During the fieldwork, I conducted participant observations and seventeen in-depth interviews with thirteen mothers and four fathers of children affected by cancer on how they had migrated for diagnosis and medical treatment for their children. Interviews were carried out in these three main areas, i.e. the hospital, the NGO and the hotel, and were recorded and transcribed respecting the anonymity and confidentiality of participants. The narrative analysis was carried out identifying codes and nodes and through the elaboration of comparative tables.

The chapter is organized in five sections, in addition to the introduction and conclusions. First, I summarize the theoretical framework in order to situate my approach. Then, I show that disruption that gives rise to a liminal phase corresponds to both, migration and disease. In the third section, I bring some characteristics of the liminal period framed in its migratory and hospital context. Then I show how liminality is prolonged by structural factors that make it impossible to return to the place of origin. Finally, in the discussion I return to the arguments supported throughout the text, articulating the analytical categories with mind the findings.

About Illness Experience

As some scholars have pointed out (Csordas 1994; Jackson 2000; Good 2003), diseases should be studied as bodily, intersubjective, emotional, symbolic and cultural experiences that acquire meanings according to different contexts and situations where they are framed. Social structures have implications on the intersubjective illness experience, as expressions of social inequalities linked to economic forces (Castro and Farmer 2004). This is related to 'what political, economic and institutional powers do to people and, reciprocally, how these forms influence responses to social problems' (Kleinman et al. 1997, 9). According to these perspectives, every analysis of the experience of suffering must include the multiple forms of violence as well as social, political and economic inequalities that characterize our societies (Grimberg 2003; Margulies 2008). In this line, many authors have argued that personal stories with disease speak about deeper aspects such as those related to lack of prevention policies and obtaining diagnoses, all of which reflects contexts of social inequality, decrease in the State's regulatory capacity in the area of health (Margulies et al. 2006) and the deterioration of health institutions (Epele 2010). In short, as Grimberg (2003) pointed out, experience constitutes a 'tense unity between action and symbolization (…) it is an active intersubjective construction, subject at the same time to historical structural processes' (81–82). So, its understanding must necessarily refer to the articulation between relationships of power, body, gender, emotions, reflexivity, in different areas of daily life.

Taking this into consideration, illness experience is 'a variable process, located in a network of intersubjective relationships, built and reconstructed historically and socially in a diversity of dimensions (cognitive, normative-evaluative, emotive, etc.)' (Grimberg 2003, 81). In this way, by studying the experience of the illness, it is possible to understand the conditions in which people experience health and disease, that is, the specificity of political and economic circumstances through which lives acquire form and constitute new subjectivities coloured by social injustice (Das and Das 2007).

From a phenomenological approach, disease has a narrative structure (Good 2003) so then, it should be studied from the stories that subjects elaborate in relation to their suffering (Garro 1994). Several authors (Heidegger 1985 in Rabelo and Souza 2003) have argued that experience is not an isolated phenomenon from a person's life. On the contrary, it maintains an organic relationship with life. An experience, say Rabelo and Souza, has a hermeneutical structure that implies a reflexive regression operation in which a previous configuration of meaning is resumed. This operation involves both understanding and affectivity and is configured according to a sense of belonging and familiarity with the social environment in which everyday actions and projects take place. In this way, every experience has a temporal order, that is, the events that happen in the life of a person are only sequentially ordered by the one who experienced them at the time of narrating them, so that the narratives, understood as 'instances of action' (Alves and Souza 1999), are the gateway to the subject's experience. In the context of the disease, the flexibility of narrative creation allows the sick person and their caregivers to cope with the biographical alteration that may have been caused by the discovery of a medical condition, make sense of their lives as sick or caregivers and (re) build their plans for the future (Kleinman 1988; Del-Vecchio Good et al. 1994; Little et al. 1998).

Immobility and Uncertainty: Fundamental Categories to Understand Cancer Care

Cancer shares a series of characteristics with other types of chronic diseases such as its prolongation over time, uncertainty regarding its course, development and cure (Glaser and Strauss 1967), being intrusive, modifying patients' lives, having a high cost (Strauss et al. 1982) and generating chronic forms of care (Mol 2008). It is recognized that the diagnosis produces a biographical disruption (Bury 1982) that affects, simultaneously, the whole system of values and meanings that sustain a daily life person (Corbin and Strauss 1985), mobilizing, in turn, the whole group broader social network that accompanies it (Luxardo and Alonso 2009).

Social sciences regarding this disease have mostly focused on adult populations (Sontag 1979; Saillant 1988; Del Vecchio Good et al. 1994;

Karakasidou 2008; Stoller 2008; Erwin 2008; Luxardo and Alonso 2009; Alonso 2009; Luxardo 2016) and many of these studies have analysed the social determinants that affect cancer care (Luxardo and Bengochea 2015), the meanings and metaphors that individuals elaborate in relation to diagnosis (Sontag 1979; Kleinman 1988), the perceptions they have from their own experience of illness (Mathews et al. 2015) as well as the meanings elaborated by professionals regarding patient attitudes (Del Vecchio Good et al. 1990), and the clinical narratives (Del Vecchio Good 1994) they elaborate.

With regard to childhood cancer, works are scarce. According to Vindrola-Padros (2011), this could be due to the social constructions on childhood, considered as a period of formation where death is not expected. In this sense, the association of cancer with death would explain, according to the author, the lack of inquiries, since the disease is constituted as a taboo. Bluebond-Langner (1978) demonstrated the ability of children to understand their disease and its management in medical terms. According to this author, the possibility of death, spatial and temporal transformations in daily life, and the decrease in social interaction contribute to the transformation of the representations that sick children have regarding their illness. Other research also identified the isolation caused by hospitalization and the biographical transition that this implies (Young et al. 2002).

Regarding migration, it is known that these displacements generate suffering and they imply multiple ways of adaptation as well as particular forms of interaction with community. Studies aligned with the 'mobility turn' agree on considering that mobility includes both a material dimension (territorial displacement) (Marzloff 2005) and imagined dimension (previous projections) related to representations about a certain trip, the purpose that it pursues, and the possibilities of its realization including the non-trip, the one that could not be realized. These perspectives presuppose a vision of the geographical transfers in material and symbolic terms that is related to the concretion of displacement as well as to the expectations of transfer and the potential trip.

As a fundamental dimension of everyday life (Salazar 2016), mobility could be understood as a resource for solving problems related to health (Vindrola-Padros et al. 2018), taking into account that this resource is unevenly distributed among people and populations (Kaufmann et al. 2004; Leivestad 2016). Not everyone who wants, imagines or wishes to mobilize is capable or able to do so. Likewise, mobility often creates or reinforces differences and inequalities (Salazar 2010). In this line, recent approaches propose to include immobility and inequality in human movement studies (Gutekunst et al. 2016), considering the restrictions imposed by systems on human mobility. In this way, Glick Schiller and Salazar (2013) proposed to understand the relationship between immobility and mobility as 'mobility regimes', referring to migratory experiences and imaginaries, emphasizing power relations and social inequalities. Hence, immobility is a constituent part of mobility. Anthropological approaches to immobility highlighted invisibilized processes in the study of contemporary transnational migration (Hannam et al. 2006; Salazar 2012). Some studies have focused on liminality as a key concept of immobility (Khan 2016, 5): 'If immobility and mobility are extremes on a continuum, what lies in-between?'

To analyse the transition that takes place between one state and another, i.e. disease-cure/death and migration-return to the place of origin, I chose the anthropological category of liminality since it allows understanding phenomena/experiences involving structural transformations, in-between situations characterized by the dislocation of established structures (Horvath et al. 2015).

Liminality in Cancer Care

Before explaining the notion of liminality that I use in this work, it is necessary to refer, briefly, to its definition in terms of how it was originally used in social anthropology. Victor Turner recovers the category of liminality formulated by Arnold Van Gennep in the book 'Rites de passage' (1909, cited in Turner 1992), to define those rites that accompany each change of place, state, social position and age. Van Gennep identified three fundamental phases of these rites of passage: separation, margin—liminality—and aggregation or incorporation. The first phase represents the moment when a status is abandoned to give rise to another. The second, limen, refers to the moment of transition and, finally, incorporation or reincorporation refers to the consummation of the rite.

Turner was interested in analysing moments of transition corresponded to the second phase delimited by Van Gennep's—liminality. Following the author, liminality refers to the margin that is generated between a past and a future structure or status, associating this with chaos, ambiguity, lack of rules and lack of structure—anti structure—resulting from the transition itself. Turner (1969) noted that liminal people or 'threshold people' have an ambiguous social status and their behaviour is normally passive, obeying instructions and accepting what is being done to them.[2] Although the concept of liminality has re-emerged via Victor Turner writings (Thomassen 2014), contemporary approaches have adopted different meanings due to the fact that this category was originally developed to analyse rituals in small-scale societies (Szakolczai 2009). Today this concept is used in many fields of knowledge and in diverse ways. Nevertheless, every approach based on liminality needs to stay close to the idea of transition. Keeping this central definition, liminality is a powerful tool of analysis to understand transformations in the contemporary world.

Some medical anthropologists have pointed out the analytical potential of liminality (Frankenberg 1986; Murphy et al. 1988; Frank 1995; Little et al. 1998; Achterberg et al. 1994; Craddock Lee 2008; Erwin 2008; Stoller 2008; Jackson 2005). In particular, some of them have argued that liminality is a fundamental category to analyse the experience of cancer (Little et al. 1998) due to the chronicity, which represents a limbo for life (Halvorson-Boyd and Hunter 1995). Other scholars (Achterberg et al. 1994) have established a certain parallelism between illness and rites of passage, in the sense that these imply a separation from the group, the dissolution of the everyday world, a time of transition—ambiguous—and the consummation of a new state (remission or death). This liminal state has also been associated with the idea of 'survivor' (Erwin 2008) and with an indeterminate state (Stoller 2008) since the diagnosis.

---

[2]A criticism of Turner's overly romantic, apolitical take on liminality and its destabilising potential has been pointed by Khan (2016).

Wainer (2015) identified that both becoming an oncology patient and undergoing medical treatment for this disease constitute transition phases. The child and his family, says the author, cross several thresholds along the therapeutic journey: from health to illness, from illness to survival or death. Even when the first cycles of chemotherapy have been passed, the child continues to go through an ambiguous phase of liminality that ranges from being sick (present) to becoming a cancer survivor (potential future). As I will show, the transition produced during medical treatment results from the combination of both migration and disease. In this way, 'potential future' it does not correspond only to becoming a cancer survivor but also to 'going back home'. Other studies pointed that hospitals are liminal spaces, 'where people are removed from their day to day lives, taken into a betwixt and between space of being diagnosed, treated, operated upon, medicated, cleansed, etc. (…) In hospitals, medical experts determine the rites of passage undertaken' (Long et al. 2008, 73).

The recently stated perspectives are appropriate to address the phenomenon of migrations produced for childhood cancer care, in the sense that this immobility process is associated with social structures and determined by local and global contexts as well as negotiations that subjects establish in their daily lives (Salazar 2020).

Cancer produces ruptures and structural changes in daily life, both of the affected people and those who accompany it and at the same time stains the life of uncertainties. In addition, the fact of having to migrate, as well as staying 'away from home' for an indeterminate period of time, imposes limitations on everything that people frequently seek to hold on to deal with the disease. The child and his family are removed from their home, routines and ties, having to overcome alone not only the difficulties that the treatment imposes on a physical and emotional level, but also the suffering caused by migration. What happens during the migratory period? How is this experience? I will try to answer these questions forward, showing how immobility and liminality represent, both, fundamental analytical categories to understand the experience of cancer and migration in Argentina.

Disruption of Everyday Life

Often when referring to cancer, the time of diagnosis is usually emphasized as a moment of rupture. However, there are multiple ruptures and mismatches in the lives of these people, as well as different ways of perceiving and experiencing such ruptures. In this regard, several authors questioned the notion of biographical disruption associated with the diagnosis by planning that every biography experiences interruptions (Estroff 1993) and that, when it comes to chronic diseases, the disease may not only be a single biographical disruption, but, 'a continuation of the disruption - one among many others' (Honkasalo 2001, 343). On the other hand, some have proposed the notion of 'biographical reinforcement' (Carricaburu and Pierret 1995), arguing that in some cases more than a rupture the disease enables to reinforce roles and bonds.

Wainer (2015) identified some stories of mothers and fathers for whom having a child with cancer represented a crisis that added to other pre-existing crises, that is, according to this author the disease, in some cases, was part of a continuum of struggles throughout life. In my own field, I could also observe that for some families the disease presented itself as a set of transformations in life and not necessarily the most significant rupture.

The diagnosis is not an isolated process and separated from the routes that preceded it (Brage 2018). On the contrary, it is inserted in previous routes that print particular senses to the experience of the disease, taking into account that, often, the correct diagnosis is something for which mothers and fathers have to fight (Vindrola-Padros and Brage 2017). Although there were cases in which the diagnosis occurred quickly and proved unexpected, these were observed to a lesser extent. A diagnosis implied comings and goings to health centres and hospitals, transferred by different provinces, among many other ruptures that 'not knowing what the kid had' implied, as a loss of employment and abandonment of activities, etc. In this sense, the participants often expressed that, although the diagnosis represented a break, in parallel they felt a sense of hope, in the sense that 'there is something to do' by clinical medicine. In other cases, the diagnosis was experienced as serious news, since knowing that the condition is chronic, not curable or of severe prognosis, it was sometimes interpreted as a sentence that will condemn the individual to live, feel and build their world in relation to this suffering.

Cancer and Migration: Disruptive Events

Following Good (2003), the everyday world of people living with chronic disease or with a potentially deadly disease—and who cares for it—is undone, deformed in the way that the normal rhythms are transformed into the rhythms of medical treatment. Thus, in the framework of hospitalization, the everyday world is replaced by the bureaucratic and clinical world as a result of routine medical practices (Del Vecchio Good et al. 1994; Good 2003), as well as the impossibility of long-term projection (Frank 1995). In this line, chronic disease, and particularly cancer, produce ruptures and structural changes in the daily life of people affected and those who accompany them. As Kleinman (1988, 20) has indicated, 'cancer is an unsettling reminder of the obdurate grain of unpredictability and uncertainty and injustice in the human condition (…)'. In addition, the fact of having to migrate, as well as staying 'away from home' for an indeterminate period of time, imposes limitations to everything that people frequently seek to cling to deal with the disease. The child and his family are far away from their home, routines and ties, having to endure in solitude not only the difficulties that the treatment imposes on a physical and emotional level, but also the suffering caused by migration and a new status of immobility.

When analysing the narratives of mothers and fathers of children with cancer, a permanent association is found between the disease and the migratory process. Both processes were narrated as disruptive and characterized by uncertainties, fears, doubts and expectations, that is to say, both events were narrated assembled, as Carolina's testimony shows:

> It is very sad, because we are so far away, but I have to be strong. Because he sees me crying and asks me: 'Mom, why do I have this disease?' he asks me and I do not know what to answer him because I have no answer. So, for him, I come for him [to Buenos Aires], because I want him to be well, to have good health, to be a healthy child and not to be worried if

something is going to happen to him, so I have to be strong. He says, 'but if I were a healthy child, we would not be here'. And I say to him, leave everything to God and the Lord will do the work. So well, that is the way we take it. I am always happy, I make him laugh so he does not get bored, because we are so far, far away. (Carolina, Aguaray, Salta)

I was entering the hotel, where I met Carolina. She and her son Kevin (11 years old) were walking out the front door into the street. Carolina had a small suitcase with wheels. When we crossed, we both greeted each other kindly. 'Hello', she said, to which I replied, 'Hello, good morning'. I sat down at the reception to talk to Glaucia, a young woman who worked at the hotel. It took about ten minutes until Carolina and Kevin appeared in the small reception. In a low voice Carolina asked Glaucia: 'Has someone called me on the phone?' To which she replied: 'As far as I know, no'. 'If someone calls me, tell them to call me in the afternoon'. 'Well', Glaucia answered again and continued asking Carolina: 'Are you leaving right now? Because here, the lady -I- wants to talk with mothers that have migrated from the North' 'Ah, okay', replied Carolina, as if she had nothing better to do. We went up to her room located on the first floor of the hotel, she turned on the fan and Kevin lay down in one of the two beds separated from each other by a distance of about one metre. The room must have had an area of three metres by four. In addition to the two beds, there was a table with two small stools where we sat down to talk. Kevin spent all that time lying on the bed playing with a cell phone. Every so often Carolina made him part of the conversation: 'right, Kevin?', and the boy nodded. Because of his illness Kevin had almost no voice. When he was three years old he was diagnosed with *larynx-pulmonary condylomatosis*, a disease that involves the growth of tumours in the respiratory tract. 'It has warts inside', Carolina explained. The disease had required several operations since these tumours frequently grow back.

When they migrated to Buenos Aires in 2009, Kevin was almost four years old. They spent two years until finally they could return to their town in the province of Salta, and since then they had to travel every month to Buenos Aires. 'It's hard to travel every month, so with my sister we take turns. We help each other like that, she comes one month and another month I come', says Carolina, who also has a six-month baby.

> This time he has come with his aunt, and doctors said he has three surgeries to be done, one has been done yesterday, and the other will be done on February thirteen … and the other one in March … And in this case, the last surgery that was done on January thirteen, went wrong. He could not breathe; the operation went wrong (…) so they had to give him oxygen (…) and the doctor said he was going to try a drug treatment [chemotherapy] … to see how it works … We have to see what happens. He told us to wait for him [kid] to be a teenager, because they say that sometimes the disease stops, it could stops growing. (Carolina, Aguaray, Salta)

The case of Carolina reflects a characteristic of this type of suffering referred to the alterations in daily life that the disease imposes. Her narrative illustrates the idea that disease and migration were combined in a 'unit of experience' (Turner and Bruner 1986, 39), which corresponds to a prolonged period of ambiguity, uncertainty and separation from the place of origin. Migration and illness are both critical and traumatic events that generate suffering but, as Carolina's narrative illustrates, both events were narrated together. Therefore, the analysis should not split them, as they constitute an assembly of disruptive events that disorganize not only the routine and the course of life, but the existential horizon of the child and his family.

On the one hand, migration produces a radical spatial-temporal change in the areas in which subjects habitually deploy their lives. On the other, the disease requires subjective and intersubjective redefinitions where new living conditions are mediated by interactions with health institutions and professionals, and the activities that organize daily life are related to bureaucracies, medicines and care required. The liminal period results then from the combination of these two disruptive events. It is a transition between a structure of past and future time, caused by both, disease and migration.

Migration does not only imply a mere geographical transfer, but it is a complex process that also involves a series of interacting factors. For those families that have had to overcome all kinds of obstacles along the way, migration represented a certain 'relief'. However, this relief was fleeting, since the arrival implies navigating between hospital and state bureaucracies, managing disease at home, getting appointments, requesting subsidies, carrying out the procedures of the drug bank, etc. In addition, at the moment people embark on the trip, they do not know how long this trip will last and what will be the different problems that they will have to deal with during this uncertain period. In this context, life starts to be governed by therapeutic rhythms (Good 2003) as well as space restricted to the hospital, all of which take place in a migratory context. This period is characterized by 'being out of time', 'not being here or there', pain, fear of death and suffering, loss of ties and social isolation, bodily, social, subjective and spatial-temporal changes, among other aspects.

Living in 'Limbo'

I observed that for mothers and fathers both events entailed spatial-temporal reconfigurations, combined with the unpredictability of the development of the disease, as well as the ever-present possibility of death. As they were outside the previous status that constituted their daily life, and separated from their primary ties, their daily environment was reconfigured in new dynamics and social relations. At the same time, the condition of immobility imposed by migration and the impossibility of returning home increased the ambiguity and isolation.

This is what I call liminal experiences, related to the transformations occurred during the medical treatment due to illness and immobility. During this liminal period, the perceptions of time and space turn out to be structured around the hospital routine, which become the organizing axis of their new daily life. Some of these transformations were not associated from the beginning to a prolonged period, but it was after months passed that this experience was consolidated and strengthened. Time and space perceptions started to be restricted and conditioned by disease, treatment and medical decisions. As one mother told me: 'we go from the hospital to the hotel and from the hotel to the hospital', referring to the impossibility of do anything outside the hospital dynamics.

By becoming the place of daily circulation, a place where life and social relationships are structured, the hospital acquired a multiplicity of meanings for these people. On several occasions, mothers and fathers narrated their need to distance themselves from the hospital, associated with confinement, stress and suffering. In the words of one of the parents: 'When we go to the hospital it is one thing and when we are outside we try not to name the hospital. We try to forget everything a bit'.

Unable to control time, space and mobility, this was often narrated as lethargic, as can be seen in the narrative of one mother: 'the treatment lasts two years, and we have to be here at least until they tell us… also we are waiting for the marrow transplant'. All this was configured in relation to the uncertain permanence in Buenos Aires. As one of the mothers explained to me: 'the treatment is long, they [doctors] suggested that we should already be thinking that it will take at least two years'. When asked about what this generated, she answered the following: 'Well … anguish, I am anguished because I am alone here… but we have to wait, there is nothing I can do (…)' (Mother, Santo Tomé, Corrientes). In this way, beyond the immediacy that some stages of the treatment imply, one of the characteristics that emanated in relation to the temporary transformations referred to the waiting, making difficult the long-term projection since the time was suspended, determined by medical information and results. All these were narrated as generating a feeling of 'impotence' by not being able to 'control' the course of disease and being forced to 'accept the new reality'.

Regarding the perception of space, this also meant a radical transformation since the circulation of these subjects, both in the hospital and in the city, was restricted to therapeutic rhythms and activities, as well as to the management of housing and the drugs. The hospital as a stage for the deployment of practices and sociocultural dynamics, that is, the place practiced (De Certeau 1996), became the place where the daily life of children and their families unfolds, as mentioned by one of the mothers: 'almost nothing we have known [of Buenos Aires] because she [daughter] is here [hospital] all day'. In this way, immobility does not only refer to the impossibility of returning home but to the limitation of mobility around the city due to hospitalization and lack of resources.

This particular moment of medical treatment, then, combines the uncertainty that characterizes every chronic disease and the suffering related to the fact of being far away from home for an indefinite period of time, as well as the impossibility to return to the place of origin and the limitation imposed over daily life motility.

This liminal period is combined with immobility, all of what was narrated as a 'loss of control' which, at the time, gives rise to different strategies to affront suffering. In these contexts, I observed that subjects often sought to restore dissolved order, even when the present was deeply ambiguous. These subjects sought to reverse and/or reduce situations of pain and suffering. This last point contrasts with Turner's (1992) perspective on liminal people as if they were 'passive'. Instead, ethnography data, as shown by Khan (2016), allows making visible some hidden aspects.

Immobility also appears as relevant when analysing the desire and contradictions of going 'back home'. Most parents expressed the desire to return to their places of origin, which contrasted with the possibilities going back home. In their narratives, the expectations of going back home were intermingled with biomedical decisions as well as contradictions and fears regarding how they felt. The contradictions were expressed in a desire to return home contrasted with the attention received in those places, expressed mostly in narratives such as the following:

- I will not go back to Corrientes until she is cured
- Why?
- In Corrientes they treated her badly...
- In which way did they treat you badly?
- They talked to me badly, they explained everything wrong and they shook her everywhere. They did not find the vein and, you're supposed to be a doctor, you have to know where the vein is! They pricked her a lot of times and there she got bruised. I arrived here [Hospital in Buenos Aires] and they punctured her once and immediately the chemo began... If I have to return to Corrientes to the follow up, well, I'm not going. I do not care if I have to pay, but I do not go back there I told the doctor (Ana, Corrientes).

As is evident in the case of Ana, the contrast between the care received in her province the one that received in the CABA was decisive in her decision of not to return until her daughter was cured. Another mother expressed the following:

> When I leave to my province, I will be terrified of going to the paediatrician because I will not feel safe… most of all I will be not feel safe at the hospital, because to the house, yes, of course I want to go back [laughs], but not to the hospital. (Sofia, El Dorado, Misiones)

These narratives illustrate contexts and realities in the provinces of origin of these people related to the unequal distribution of health services and more qualified professionals throughout the country, making impossible to return home, even when medical treatment is ended and required just follow up. These aspects I have pointed out summarize briefly the reality in Argentina, making evident the different ways in which immobility is related to structural process that could increase and prolong suffering. However, it is not only the unequal distribution of health services which influence mobility but, above all, the possibilities and resources that people are able to take advantage of. Conceived from this perspective, immobility is determined not only by the spatial structure and transport infrastructure, but also by the skills, knowledge and possibilities of people themselves, as part of the social capital of groups and individuals.

Discussion

As I tried to show, immobility and liminality are fundamental analytical categories to understand the experience of cancer and migration in Argentina. Far from being a secondary aspect to the experience of the disease, immobility constitutes a central event within the therapeutic itineraries of childhood cancer (Brage 2018). Moving, as well as not been able to move is decisive in the course of disease and the therapeutic results because if subjects migrate they have greater healing possibilities as well as access to better treatments. On the other hand, liminality in the case of children affected by cancer and their families is not something that 'just happens' because of the fact of being affected by cancer. Instead, it is a state of ambiguity that it is exacerbated by the impossibility of returning home, related to lack of health policies, disparities in health access and social inequalities, unequal distribution of health services, and lack of resources in the north of the country.

Analysing illness experience through these perspectives allowed me to go beyond medical practice and clinical explanations on cancer. Instead, I tried to focus on bringing together clinical moments with daily life. While in medicine, uncertainty is often theorized in relation to clinical care and medical technologies, medical anthropology has improved these approaches by taking in consideration the health/disease/care process which wraps up not only healthcare services, but many other ways of solving and dealing with the disease, including the family, the community and the closest social networks. Paying attention on other situations that occur within the daily life of patients and their caregivers, beyond the medical sphere, brings out new possibilities on interpreting uncertainty. This article has shown that events are not segregated when people narrated their experiences. In this way, I tried to conceptualize illness experience beyond the individual conception of body, showing how suffering is experience in a complex intersubjective network.

Illness as well as the migratory process implied for mothers, fathers and children the dissolution of the everyday world and the beginning of a new period characterized by uncertainty. Since they arrived at Buenos Aires they entered a liminal state characterized by ambiguity, migratory suffering and the impossibility of making long-term plans. Both events produce suffering that it is not only related, as I recently mention, to the fact of being affected by cancer. Instead, the basis of this suffering is founded in social structures, which through different forms of inequality produce different ways of living, sickening, healing and dying, all of which shows that mobilizing geographically, as well as not being able to do so, is, then, result of social, cultural, economic and political negotiations (Sheller and Urry 2006; Urry 2007).

The phase initiated with illness and the migratory process give rise to an uncertain moment, which I called, liminality, an in-between in the process of mobility and immobility, as well as a process of health/illness. This liminal phase has an intrinsic relation with immobility in the sense that migration is prolonged by structural factors as well as medical decisions that make not possible to return home. This last is also combined with the ambiguity and uncertainty of cancer. Life changed absolutely because of disease and migration and, as both events were narrated together and linked, it is not possible to isolate each other, since they constitute the illness experience.

Carolina's stories about her experience with her son's illness show that there is no division between the subject and the social networks in which he is ingrained. Nor is there a fragmentation of everyday life with the disease. Rather, what occur during this period are spatio-temporal transformations that print different senses to the experience of the disease. In this way, the analytical richness of these narratives shows that the ambiguity and the state of liminality that people live is structured, not only in terms of a disease conceived in biological terms, but also in terms of structural conditions that increase or decrease the uncertainties

Conclusions

As this paper tried to show, immobility is central when analysing cancer medical care in Argentina. Long far of being isolated aspects, migration and disease constitute a unit of experience, framed in particular contexts. In this way, the paper tried to argue that transformations produced by migration and hospitalization as a daily life are intrinsically linked. The impossibility of returning home (immobility) configures a prolonged liminal state which, at the same time, it is combined with the uncertainty produced by cancer.

References

Abriata, M. Graciela, and Florencia Moreno. 2010. "Cáncer en la Población de Menores de 15 Años en Argentina." *Rev Argent Salud Pública* 1 (3): 42–45.

Achterberg, Jeanne, Barbara Dossey, and Leslie Kolkmeier. 1994. *Rituals of Healing: Using Imagery for Health and Wellness*. New York: Bantam Doubleday Dell Pub.

Alonso, Juan Pedro. 2009. El cuerpo hipervigilado: incertidumbre y corporalidad en la experiencia de la enfermedad en Cuidados Paliativos. In *Cuadernos de Antropología Social FFyL-UBA*, pp. 103–120.

Alves, Paulo Cesar, and Iara Maria Souza. 1999. Escolha e Avaliação de Tratamento para Problemas de Saúde: considerações sobre o itinerário terapêutico en Rabelo, MCM., Alves, PCB., and Souza, IMA. *Experiência de doença e narrativa*. Rio de Janeiro: Editora FIOCRUZ.

Aurelie Pujol, C. J, C. L. Bertone, and L. D. Acosta. 2014. "Morbidity and Mortality Rates for Childhood Cancer in Argentina. 2006–2008." *Arch Argent Pediatr* 112 (1): 50–54.

Bury, Michael. 1982. "Chronic Illness as Biographical Disruption." *Sociology of Health & Illness* 4 (2): 167–182.

Bluebond-Langner, Myra. 1978. *The Private Worlds of Dying Children*. Princeton: Princeton University Press.

Brage, Eugenia. 2018. *"Si no fuera porque me vine…"*: Itinerarios terapéuticos y prácticas de cuidado en el marco de las migraciones producidas desde el Noroeste y Noreste Argentino para la atención del cáncer infantil. Un abordaje antropológico. Tesis de doctorado. Facultad de Filosofía y Letras, Universidad de Buenos Aires.

Carricaburu, Danièle and Janine Pierret. 1995. "From Biographical Disruption to Biographical Reinforcement: The Case of HIV-Positive Men." *Sociology of Health & Illness*. https://doi.org/10.1111/1467-9566.ep10934486.

Castro, Arachu, and Paul Farmer. 2004. "Pearls of the Antilles? Public Health in Haiti and Cuba." In *Unhealthy Health Policy: A Critical Anthropological Examination*, edited by A. Castro and M. Singer.

Corbin, Juliet, and Anselm L. Strauss. 1985. "Managing Chronic Illness at Home: Three Lines of Work." *Qualitative Sociology* 8: 224–247.

Craddock Lee, Simon. 2008. "Notes from White Flint. Identity, Ambiguity and Disparities in Cancer." In *Confronting Cancer: Metaphors, Advocacy and Anthropology*, edited by Juliet McMullin and Diane Weiner, 165–186. Santa Fe: School for Advanced Research Press.

Csordas, Thomas. 1994. *Embodiment and Experience: The Existential Ground of Culture and Self*. Cambridge: Cambridge University Press.

Das, Veena, and Rehendra Das. 2007. "How the Body Speaks: Illness and the Lifeworld Among the Urban Poor." In *Subjectivity Ethnographic Investigations*, edited by Joao Biehl, Byron Good, and Arthur Kleinman. Berkeley, CA: University of California Press.

De Certeau, Michell. 1996. *A Invencao do Cotidiano: Artes de Fazer*. 2 ed. Petropolis: Vozes.

Del Vecchio Good, Mary-Jo, Byron J. Good, Cynthia Schaffer, and Stuart E. Lind. 1990. "American Oncology and the Discourse on Hope." *Culture, Medicine and Psychiatry* 14 (1): 59–79.

Del Vecchio Good, Mary-Jo, Tseunetsugu Munakata, Yasuki Kobayashi, Cherryl Mattingly, and Byron Good. 1994. "Oncology and Narrative Time." *Social Science and Medicine* 38 (6): 855–862.

Direccion de Estadisticas e Informacion de Salud: 'Principales causas de defunción según grupos de edad' en Estadísticas vitales Información Básica - 2010 Serie 5 - Nro. 54, Buenos Aires, Argentina, Dic 2011.

Epele, Maria. 2010. *Sujetar por la Herida. Una Etnografía sobre Drogas, Pobreza y Salud*, 23–34. Buenos Aires: Paidós.

Erwin, Deborah. 2008. "The Witness Project: Narratives That Shape the Cancer Experience for African American Women." In *Confronting Cancer: Metaphors, Advocacy, and Anthropology*, edited by J. McMullin and D. Weiner, 125–146. Región Centro: School of American Research Press.

Estroff, Sue E. 1993. "Identity, Disability, and Schizophrenia: The Problem of Chronicity." In *Knowledge, Power, and Practice: The Anthropology of Medicine and Everyday life*, edited by S. Lindenbaum and M. Lock, 247–286. Berkeley, CA: University of California Press.

Farmer, Paul. 2007. "Una Antropología de la Violencia Estructural. El Caso de Haití." *Temas* 52: 63, 73.

Felice, Maria, Veronica Díaz, Vanina Livio, Mercedes García Domínguez, Liliana Franco, Eugenia Ensabella Ramos, and Juan Chaín. 2013. "Análisis de la mortalidad en enfermedades hemato-oncológicas malignas en pediatría enhospitales públicos de Argentina." *Revista Argentina Salud Pública* 4 (14): 23–31.

Frank, Arthur. 1995. *The Wounded Storyteller: Body, Illness, and Ethics*. Chicago: University of Chicago Press.

Frankenberg, Ronald. 1986. "Sickness as Cultural Performance: Drama, Trajectory, and Pilgrimage. Root Metaphors and the Making Social of Disease." *International Journal of Health Services* 16 (4): 603–626.

Garro, Linda C. 1994. "Narrative Representations of Chronic Illness Experience: Cultural Models of Illness, Mind, and Body in Stories Concerning the Temporomandibular Joint (TMJ)." *Social Science & Medicine* 38 (6): 775–788.

Glaser, Barney G., and Anselm L. Strauss. 1967. *The Discovery of Grounded Theory: Strategies for Qualitative Research*. Chicago: Aldine Publishing Company.

Glick Schiller, Nina, and Noel Salazar. 2013. "Regimes of Mobility Across the Globe." *Journal of Ethnic and Migration Studies* 39 (2): 183–200.

Good, Byron. 2003. *Medicina, Racionalidad y Experiencia. Una Perspectiva Antropolgógica*. Ediciones Bellaterra. Barcelona.

Grimberg, Mabel. 2003. "Narrativas del Cuerpo: Experiencia Cotidiana y Género en Personas que Viven con VIH." *Cuadernos de Antropología social* 17: 79–99.

Gutekunst, Miriam, Andreas Hackl, Sabina Leoncini, Julia Sofia Schwarz, and Irene Götz, eds. 2016. *Bounded Mobilities: Ethnographic Perspectives on Social Hierarchies and Global Inequalities*. Bielefeld: Transcript Verlag.

Gutiérrez, Andrea. 2008. "Geografía, Transporte y Movilidad." *Espacios Geografía* 37: 100–107.

Halvorson-Boyd, Glenna, and Lisa Hunter. 1995. *Dancing in Limbo: Making Sense of Life After Cancer*. Wiley, ISBN 0787901032, 9780787901035. Length 192 pages.

Hannam, Kevin, Mimi Sheller, and John Urry. 2006. "Mobilities, Immobilities, and Moorings." *Mobilities* 1: 1–22.

Honkasalo, Marja Liisa. 2001. "Vicissitudes of Pain and Suffering: Chronic Pain and Liminality." *Medical Anthropology* 19 (4) (January): 319–353. https://doi.org/10.1080/01459740.2001.9966181.

Horvath, Agnes, Bjørn Thomassen, and Harald Wydra. 2015. *Breaking Boundaries: Varieties of Liminality*. Berghahn Books.

Instituto Nacional del Cáncer, Ministerio de Salud de la Nación. n.d. https://www.argentina.gob.ar/salud/inc.

Jackson, Jane. 2000 *Camp Pain Talking with Chronic Pain Patients*. Philadelphia: University of Pennsylvania Press.

Jackson, Jean. 2005. "Stigma, Liminality and Chronic Pain: Mind-Body Borderlands." *American Ethnologist* 32 (3): 332–353.

Karakasidou, Anastasia 2008. "The Elusive Subversion of Order: Cancer in Modern Creete, Greece." In *Confronting Cancer: Metaphors, Advocacy, and Anthropology*, edited by J. McMullin and D. Weiner. Región Centro: School of American Research Press.

Kaufmann, Vincent, Manfred Max Bergman, and Dominique Joye. 2004. "Motility: Mobility as Capital." *International Journal of Urban and Regional Research* 28 (4): 745–756.

Khan, Nichola. 2016. *Keywords of Mobility: Critical Engagements*. Edited by Noel Salazar and Kiran Jayaram. New York and Oxford: Berghahn Books, Chapter 5.

Kleinman, Arthur. 1988. *The Illness Narratives: Suffering, Healing, and the Human*. Condition. New York: Basic Books.

Kleinman, Arthur, Veena Das, and Margaret Lock. 1997. *Social Suffering*. Berkeley, CA: University of California Press.

Leivestad, Hege Høyer. 2016. "Motility." In *The Keywords of Mobility: Critical Engagements*, edited by Noel B. Salazar and Kiran Jayaram. Oxford: Berghahn.

Little, M., C. F. Jordens, K. Paul, K. Montgomery, and B. Philipson. 1998. "Liminality: A Major Category of the Experience of Cancer Illness." *Social Science & Medicine 47* (10): 1485–1494.

Long, Debbi, Cynthia Hunter, and Sjaak van der Geest. 2008. "When the Field Is a Ward or a Clinic: Hospital Ethnography." *Anthropology and Medicine* 15 (2): 71–78.

Luxardo, Natalia 2016. "Morir en la Propia Ley. Lógicas y Supuestos Permeando Evaluaciones de Proyectos de Investigación de Ciencias Sociales en el Campo de la Salud." *Revista Debate Público. Reflexión de Trabajo Social* 6 (12): 43–58.

Luxardo, Natalia, and J. P. Alonso. 2009. Cáncer e identidades en el final de la vida Scripta Ethnologica, núm XXXI, 2009, pp. 17–32. Consejo Nacional de Investigaciones Científicas y Técnicas Buenos Aires, Argentina.

Luxardo, Natalia, and Laura Bengochea [Comps]. 2015 *Cáncer y Sociedad. Aportes desde Múltiples Disciplinas*. Buenos Aires: Editorial Biblos.

Margulies, Susana. 2008 Construcción social y VIH/sida. Los Procesos de Atención Médica. Tesis de Doctorado, Facultad de Filosofía y Letras, Universidad de Buenos Aires.

Margulies, Susana, Nelida Barber, and Maria Laura Recoder. 2006. "VIH SIDA y 'Adherencia' al tratamiento. Enfoques y perspectivas." *Antípoda. Revista de Antropología y Arqueología*.

Marzloff, Bruno. 2005. *Mobilités, Trajectoires Fluides*. La Tour d'Aigues: Editions de l'Aube.

Mathews, Holly, Nancy Burke, and Eirini Kampriani. 2015. *Anthropologies of Cancer in Transnational Worlds*. New York: Routledge.

Mol, Annemarie. 2008. *The Logic of Care: Health and Problem of Patient Choice.* London: Routledge.

Moreno, Florencia, Veronica Dussel, and Liliana Orellana. 2015. "Childhood Cancer in Argentina: Survival 2000–2007." *Cancer Epidemiology* 39 (4): 505–510. https://doi.org/10.1016/j.canep.2015.04.010. Epub 2015 May 18.

Murphy, R. F., J. Scheer, Y. Murphy, and R. Mack. 1988. "Physical Disability and Social Liminality: A Study in the Rituals of Adversity." *Social Science & Medicine* 26 (2): 235–242.

Rabelo, Miriam, and Iara Souza. 2003. "On the Meaning of Nervoso in the Trajectory of Urban Working-Class Women in Northeast Brazil." *Ethnography* 4 (3): 333–361.

Registro Oncopediátrico Hospitalario Argentino (ROHA). 2012. Incidencia 2000–2009 supervivencia 2000–2007 tendencia de mortalidad 1997–2010 / Florencia Moreno, et al. - 1a ed. - Buenos Aires: Instituto Nacional del Cáncer, Ministerio de Salud de la Nación.

Saillant, Francine 1988. *Cancer et Culture: Produir le Sens de la Maladie.* Montreal, Editions Saint Martin.

Salazar, Noel B. 2010. "Towards an Anthropology of Cultural Mobilities." *Crossings: Journal of Migration and Culture* 1 (1): 53–68.

Salazar, Noel B. 2012. "Imagining (Im)mobility at the 'End of the World.'" In *Technologies of Mobility in the Americas*, edited by Phillip Vannini, Paola Jiron, Ole Jensen, Lucy Budd, and Christian Fisker. New York: Peter Lang.

Salazar, Noel B. 2016. "Keywords of Mobility: What's in a Name?" In *Keywords of Mobility: Critical Engagements,* edited by N. Salazar and K. Jayaram, pp. 1–12. Oxford, UK: Berghahn Books.

Salazar, Noel. B. 2020. "On Imagination and Imaginaries, Mobility and Immobility: Seeing the Forest for the Trees." *Culture & Psychology.*

Sheller, Mimi and Urry, John. 2006. "The New Mobilities Paradigm." *Environment and Planning* 38: 207–226.

Sontag, Susan. 1979. *Illness as Metaphor.* New York: Vintage.

Stoller, Paul. 2008. "Remissioning Life, Reconfiguring Anthropology." In *Confronting Cancer: Metaphors, Advocacy and Anthropology,* edited by Juliet McMullin and Diane Weiner, 27–42. Santa Fe: School for Advanced Research Press.

Strauss, Anselm L., Shizuko Fagerhaugh, Barbara Suczek, and Carolyn Wiener. 1982. "The Work of Hospitalized Patients." *Social Science and Medicine* 16 (9): 977–986.

Szakolczai, Arpad. 2009. "Liminality and Experience: Structuring Transitory Situations and Transformative Events." *International Political Anthropology.*
Thomassen, Bjørn. 2014. *Liminality and the Modern: Living Through the In-Between.* 1st ed. London: Routledge.
Turner, Victor. 1992. "Liminality and Communitas." In *The Ritual Process: Structure and Anti-Structure.* Chicago: Alpine Publishing, 1969.
Turner, Victor, and E. Bruner. 1986. *The Anthropology of Experience.* Urbana: University of Illinois Press.
Urry, John. 2007. *Mobilities.* Cambridge, UK: Polity Press.
Vindrola-Padros, Cecilia. 2011. "The Disease and Treatment." *Children and Society* 26 (6): 430–442. "Life and Death Journeys: Medical Travel, Cancer, and Children in Argentina." Graduate School Theses and Dissertations. Paper 3395. http://scholarcommons.usf.edu/etd/3395: University of South Florida.
Vindrola-Padros, Cecilia, and Eugenia Brage. 2017. "Child Medical Travel in Argentina: Narratives of Family Separation and Moving Away from Home." In *Children's Health and Wellbeing in Urban Environments*, edited by C. Ergler, R. Kearns, and K. Witten, 128–144. London: Routledge.
Vindrola-Padros, Cecilia, Ginger A. Johnson, and A. Pfister. 2018. *Healthcare in Motion Immobilities in Health Service Delivery and Access.* New York and Oxford: Berghahn Books.
Wainer, Rafael. 2015. *Permeable Bodies: Children, Cancer, and Biomedicine in Argentina.* The University of British Columbia.
Young, Bridget, Mary Dixon-Woods, Michelle Findlay, and David Heney. 2002. "Parenting in a Crisis: Conceptualising Mothers of Children with Cancer." *Social Science & Medicine* 55 (10): 1835–1847. https://doi.org/10.1016/s0277-9536(01)00318-5.

Part II

Embodied Stillness and Fixity

Part II

Embodied Softness and Body

6

Embodying Immobility: Dysphoric Geographies of Labour Migration and Their Transformations in the Therapeutic Context of 'Venda' Ancestor Possession in Post-apartheid South Africa

Vendula Rezacova

Introduction

Tshele,[1] the first phase of ancestor possession undertaken 'sitting down' (*ngoma yo dzula*), was being held in May 2006 in the household of one of the initiated 'daughters' of the female healer with whom I lived, in a remote, hardly accessible settlement of Ha-Mashau southwest of the local urban centre, Thohoyandou.[2] An exceptionally high level of agricultural activity for the parched region was visible, goats standing atop hut-plinths and small corn fields surrounding rondavels and houses, although the small produce was sold on the market and was

---

[1] Literally 'rattles', referencing the musical instruments which are used to denote possession rituals undertaken in a posture of kneeling on the ground of a hut or house.

[2] This chapter is based on fieldwork conducted in different locations within 'Venda' (Vhembe District), Limpopo Province, and to a lesser extent Gauteng (Johannesburg and Pretoria), South Africa, between 2004 and 2006.

V. Rezacova (✉)
Institute of Sociological Studies, Charles University, Prague, Czech Republic

completed by migrants' earnings. Both the main healer, Unice, an energetic woman in her late 50s, and the spirit-afflicted 34-year-old man, Tshimangadzo, worked as labour migrants in *Tshikuani*, literally 'place of white ways/things'—referring to the Gauteng urban vortex mostly comprising Johannesburg and Pretoria. A seamstress and a miner, both became 'sick', *u lwala*, after several years of working in *Tshikuani*, to the point of 'not being able to move from bed'. They also experienced disruptions of work technologies in *Tshikuani*—of the sewing machines and identity cards—unexplainable to technicians and therefore an unequivocal sign that their ancestor spirits, *midzimu*, were definitely interfering in the migrants' well-being. Unice had already settled as a healer of some repute in her mother's household in 'Venda', despite the latter's protestations, to abide by the ancestors' wishes that she leaves *Tshikuani* and the day job to regain health. Tshimangadzo intimated he did not have much choice but to follow suit if he wanted to live.

A number of participants gathered before sunset on the winter Saturday, including Tshimangadzo's sister, driving from *Tshikuani* picking up their mother and a number of elder female relatives and acquaintances from different parts of 'Venda' on the way; Unice's younger brother; and a number of predominantly female neighbours. The proceedings began with Unice treating the rooms of the house and the yard with protective medicines, *mushonga*, mixed with her own urine, in case evil-intentioned people or witches, *vhaloi*, tried to interfere with the possession process. Two somewhat drunken neighbours who tried to play the rattles were chased away. The playing of rattles and singing gradually picked up on tempo, as did the strenuous movements of Tshimangadzo kneeling on the floor animating his torso. Although he previously complained the gestures of *tshele* were too 'feminine' and strenuous for him to command, eventually two ancestor spirits—male and female—'came', *u swika*, to dance and communicate through his body and in a respectable voice. The proceedings were a success, and the spirits have spoken clearly as to their wishes of having Tshimangadzo stay in 'Venda' and undergo 'traditional' healer's initiation (*u twasa*), and of his relatives acquiring more ancestral cloths: bought from the Indian-owned shop in Thohoyandou. That was also the message Tshimangadzo's mother jubilantly communicated to her elder sister over the mobile phone the next morning, although she earlier confessed she was not happy about Tshimangadzo's leaving *Tshikuani* for good. The women neighbours put their slip-ons and Nike-imitation hats back on and shuffled off to attend Sunday church service. Tshimangadzo's sister departed in her car back to *Tshikuani*, dropping people off along the way. Tshimangadzo could not recall anything from the previous night—was he not sleeping?

Immobility as Culturally Mediated Praxis

The present account of therapeutically mediated immobility aims to complement analyses which have interpreted the incapability or unwillingness to move from place or 'move on' of particular subjects—individuals and groups—as culturally creative praxis (Khan 2013, 2016). As the case study has shown, immobility associated with 'Venda' ancestor possession in the specific South African setting may be regarded equivocally by those whom such 'diagnosis' concerns. Rather than pathological states of social and existential 'stuckedness' (Khan 2016, 100–104), however, immobility—or severely restricted mobility—has constituted a culturally and personally meaningful condition when mediated by ancestor spirit possession. It has been associated with expanded personal powers, knowledge and command of one's own life by persons who have undergone its therapeutic procedures, including returning to a 'rural' area within 'Venda'.[3] The task of this chapter is to contextualize such complex sociocultural processes of harnessing health and well-being for self and others as ways of negotiating historical transformations engendered in apartheid rule, resonating in symbolic practices into the sociocultural conjunctures of post-apartheid South Africa.

---

[3]The region commonly denoted as 'Venda' was constituted as a 'homeland' in 1962 as part of apartheid's stepping up of the policy of racial segregation. A large section of the Tshivenda-speaking population was constrained to this territory. The policy of the homelands was abandoned in 1994; the territory of 'Venda' was incorporated into the Vhembe District, Limpopo Province. The territory of Vhembe corresponds with that of the former 'Venda' homeland, apart from also incorporating two areas of the former Gazankulu.

An important lens onto ancestor possession as culturally creative praxis articulating concerns with 'immobility' has been underpinned by the geographical focus of possession syndromes and of the therapeutic procedures aiming to redress them. Only 'rural' locations within the territory of the former 'Venda homeland' have been rendered conducive to the organizing of the dramatic events of possession rituals (*tshele* and *malombo*[4]). These healing rituals could not be conducted in *Tshikuani*, nor any other location within South Africa and beyond, while participants have had to travel from different locations within South Africa to attend them. As the explanation for this state of affairs was put to me by my informants: 'Ancestors would refuse to arrive, *swika*, (elsewhere)'. Moreover, as a matter of 'cure' in the mid-2000s, migrants afflicted by ancestor spirits in the urban centres have had to return to a 'rural' location within 'Venda' to take up the profession as 'traditional' healers, *nanga* or *(vho-)maine*. If they ventured beyond for short periods of time, especially to *Tshikuani*, possessed persons risked having debilitating symptoms relapse. Such stress on territoriality articulated through ancestor possession has not been expressed in terms of the ideals of sedentarism according to which particular 'people' should rightly reside in 'their own' territory. Rather, it has entailed a complex series of negotiations between living humans and their ancestors whereby territorial transfer and geographically indexed immobility of possessed ex-migrants has emerged as a way of revisioning past dichotomies of the 'city' and the 'country' of apartheid South Africa in present experience and practice.

Ancestor Spirit Possession in *Tshikuani*: Beyond 'Rebellion'

That attacks of spirit possession can intervene in the (not only industrial) workplace as people have become drawn into the corporate regimes of global economy has been observed in a number of global contexts (Ong 1988; Sharp 1994). In the setting of the microchip factories in

---

[4]Malombo or 'drum of standing up' (*ngoma yo ima*) constitutes the second phase of ritual proceedings mediating ancestor possession, successive upon 'tshele'.

Malaysia, Ong has shown that spirit possession episodes overwhelming recent women migrants from the rural villages have constituted acts of rebellion against, and dramatic commentary upon, the disruptions which the shop floor has entailed for the women's bodies and 'indigenous' moral boundaries (Ong 1988, 38). Similarly, Venda rural labour migrants—both men and women—experience ancestor spirit attacks, among other symptoms of a conditions denoted as 'illness', *u lwala*, also in disruptions of work routines and technologies in the urban workplace of *Tshikuani* (Gauteng). In the Malaysian context, possession episodes had limited consequences for the women's life and work circumstances, having been contained by biomedical categories and interventions of the *bomoh* ('traditional' healer). In contrast, 'Venda' migrants are therapeutically compelled by 'traditional' healers to leave not only the workplace where possession occurred, but also the urban vortex of Gauteng itself—*Tshikuani*, 'the place of white things'. As a route to cure the migrants undergo a complex regimen of medicines and initiation to become full-time ancestral mediums, 'traditional' healers themselves. The final condition of continued well-being—so the ancestors insist—is achieved by the migrants' re-territorialization: as ancestral mediums, the migrants have to return to 'Venda' and settle in a rural household within its territorial boundaries. Venturing beyond, and especially to *Tshikuani*, they risk having debilitating symptoms recur.

If read through Ong's (1988) analytical lens of spirit possession as counter-hegemonic commentary levelled by rural people against alienation experienced in regimes of capitalist production—with the culturally mediated option of actually leaving them for good, we would be led to an impoverished account of 'Venda' ancestor possession. For unlike the Malay female workers who were the first generation of the *kampong* population to venture into the industrial workplace (Ong 1988, 33), in the context of 'Venda' in South Africa rural–urban labour migration has been part of people's lives for well over a century—longer than ancestor possession itself was part of 'Venda' therapeutic repertoire (more on this below). Moreover, labour migration was a major component of the state-imposed system of racial segregation according to which specific categories of persons—'black' Africans—were impelled to stay in the rural 'native reserve' (later 'homeland') while others were channelled to the urban centre and capitalist workplace (Wolpe 1972). Although it was a result of coercive measures of the colonial and then apartheid political economy, labour migration to Gauteng (*Tshikuani*) has been endowed with overwhelmingly positive connotations. Becoming a labour migrant capable of maintaining a job in *Tshikuani* and sending remittances to dependants in 'Venda' has historically been associated with great degree of prestige. For men in particular, this form of self-realization closely tied to obligations to stay-at-home kin through remittances has been upheld as 'our culture', *mvelele yashu*. These two aspects of migrant experience—its initially state-coerced character and its association with cultural value—must be born in mind when trying to interpret the ambivalent interventions of ancestor spirit possession in processes of labour migration and repatriation. I will try to convey some of this ambivalence as it has been articulated though '*Tshikuani* illness' afflicting Tshivenda-speaking labour migrants, and through the transformations of its therapeutic mediations as ancestor possession in different historical periods. Such an approach can help us better appreciate the sorts of concerns which have been articulated through contemporary therapeutic immobility in South Africa.

Therapeutic Articulations of the 'Denying Body' Migrants in *Tshikuani*

Tshikuani 'illness' diagnosed as ancestor possession, a phenomenon so far absent in existing ethnographic or historical accounts, has been a very marginal phenomenon until the late 1980s[5]. The first cases of *Tshikuani* illness reported by former, invariably male, migrants date back to the 1960s. This was the most intense period of the coercive apartheid-instigated system of rural–urban labour migration, which engaged primarily of able-bodied men from the 'homelands' (Wolpe 1972, 426), including 'Venda', in the Gauteng mines and industrial complexes. Some

---

[5]This section is based on life histories gathered during fieldwork in South Africa 2004–2006. Neither Stayt (1931) nor Blacking (1985) mention the phenomenon of '*Tshikuani* illness' of migrants in their ethnographies of possession phenomena in the 'Venda' cultural context.

symptoms have remained constant throughout the historical period from the 1960s to mid-2000s: loss of appetite, emaciation, shoulder and neck pain shooting into the arms,[6] dizziness, black-outs and 'night sickness', *vhulwadze zwa vhusiku* (sleeplessness and having 'dreams which are not spoken about'). However, such symptoms do not in themselves call for a consultation with a traditional healer and the diagnosis of ancestor possession involving relocation to 'Venda'. Aforementioned symptoms are usually treated by over-the-counter medicines ('pills', *dzipilisi*), in Christian churches and/or by a regimen of traditional herbal medicines, *mushonga*. Many migrants in *Tshikuani*, as well as clients of divinations in 'Venda' which I witnessed during fieldwork between 2004 and 2006, have at one point or another experienced some or all of the above-mentioned symptoms. For an illness experienced in *Tshikuani* to be divined as ancestrally induced, the migrants have undergone a further series of untoward events relating to their embodied presence in the workplace and the city life-world. Disorientation, leading to their becoming lost in the city wandering aimlessly on its streets, having visions of white goats grazing on patches of grass amid the urban sprawl, is one of them. Crucially, their 'bodies were denying', *muvhili u khou hana*, in very specific circumstances of their engagements with the urban workplace.

The 'denying body' was experienced in ways specific to the performance of habitual work routines as these have been defined by the position in the urban structure of the labour market, mapping across class and gender. Moreover, with the shift from manual to service and managerial positions for 'black' Africans enabled by the fall of apartheid in 1994 (although dissipating earlier in late 1980s, see Posel 2004), the 'denying body' has come to encompass new symptoms. The earlier generation of migrants, constrained to the lowest and most manual rungs of the urban economy, and contemporary migrants working as miners, factory and construction workers, technical personnel, usually complained of 'arms denying'—to pick up the shovel, engage with assembly line, with the sewing machine. However, the 'denying body'

---

[6]As opposed to 'pain in the stomach' which indicates the state of being a victim of witchcraft attack, vhuloi. Shoulder and neck pain indicates problems in relations with the ancestors—most frequently spirit possession.

was not an aspect of a 'somatic culture' of the weak and disenfranchised, occupying lowly economic positions of manual labour who also use their bodies to express awareness of the socio-economic and political contradictions of their worlds (Scheper-Hughes 1993, 185). Migrants in secure middle-class positions, such as managers and teachers, experienced symptoms no less somatically indexed despite their 'mindful labour': they 'lost voice' during presentations or were 'unable to stand on their legs', *ima kha milenzhe*, after they woke up in the morning. En route to work, in a train, some migrants would feel the 'spirit', *muya*, rising into their head causing pressure so that they would fly out of the window of the carriage—a hydraulic metaphor evocative of conditions of 'nerves' (Low 1994). Their fall would be restrained by fellow travellers who began uttering Christian prayers and singing hymns, recognizing the episode as needing extraordinary attention.

Despite the evocativeness of the 'denying body', this was not an idiom articulating, in a language of mute signs, a clear-cut rebellion of the migrants against the capitalist urban economy. Some of the migrants were well positioned in it and boasted of having had driven BMWs, owned large houses, enjoying the perks of the urban lifestyle. It is also striking that the afflicted migrants continued working after the onset of their symptoms for up to several years. In no sense were they eager to leave urban employment, as I witnessed and as I have been told by the migrants in their life histories. According to the migrants' biographies, the symptoms of illness worsened as they experienced conflicts with bosses, superiors and co-workers. In this sense the 'denying body' is a culturally well-articulated syndrome which also afflicts rural women who experience 'limb paralysis' in situations of household strife. In both cases, these phenomena could be glossed as instances of 'somatization'. In situations of entrenched hierarchies of power, people express distress through embodied signs when other more conventional means of communication are not available (Good and Kleinman 1985).

The conceptual framework of 'somatization', however, would leave a crucial dimension in the interpretations of these paralysing symptoms unaddressed—namely their other-than-human causation, as it is conceived in this particular cultural context. For while rural women are diagnosed and treated as being possessed by an evil spirit, *mimuya*, the migrants' equally somatically indexed experiences of disempowerment are interpreted as caused by the spirits of the ancestors. This radical bifurcation bellies any facile identification of somatization in the syndrome of the 'denying body', as this must be simultaneously appropriately culturally grounded. The case of the possessed rural women lies outside of the scope of this chapter but serves to show that identifying ancestor spirits as the causative agents behind migrants' 'limb paralysis' is a matter of important cultural distinction on the part of the therapeutic repertoire of 'traditional' medicine. The 'denying body' is not simply a process of 'somatised' disorders which could be readily medicalized (Taussig 1980), that is contained within the suffering body (or 'mind') without challenging its encompassing social context. In contrast, the 'denying body' of migrants as it is weaved through its cultural definition of ancestor possession triggers off a series of transformations on the level of both social relations and identities. Crucially, it ensues in the sufferer's radical re-territorialization: from *Tshikuani* to 'Venda'. That this re-territorialization has become a non-negotiable ancestral wish—and the precondition for cure of an otherwise near-death existence[7]—is a fairly recent development, gaining force since the late 1990s. Furthermore, only migrants suffering from the 'denying body' in *Tshikuani*, and no other South African or world city, are in this way 'repatriated' as a matter of ancestrally mediated cure. These aspects of ancestor possession of *Tshikuani* migrants will be addressed as a way of arguing for a culturally and historically specific mediation of immobility in a therapeutic context which has itself been transforming as a way of articulating changing concerns of both sufferers, healers and the wider community.

---

[7]'The ancestors don't kill you, they just make you suffer like you wish your own death'.

Historical Context: Towards *Ancestor* Possession

Possession phenomena became part of the 'Venda' therapeutic repertoire[8] during the last quarter of the nineteenth century. This is roughly in the same period as European missionary and colonial agents made major inroads into the north-eastern Transvaal where 'Venda' existed as a congeries of chiefdoms and groups of intersecting allegiance.[9] 'Venda' possession congealed as possession by exclusively human-like spirits—ancestors, *midzimu*—with known, traceable genealogical links to the ego, reaching up to five generations, sometime during the period around World War II.[10] Before this solidification of 'ancestor possession cult', de-centralized, individual possession by foreign spirits lodged in treetops or by the spirit of the snake, as well as by ancestors 'absurdly remote' (Stayt 1931, 302) was equally possible. The decade between 1940s and 1950s was a historical period of far-reaching socio-economic and political transformations in the region of north-eastern Transvaal ('Venda'). One of these, addressed through the emergence of ancestor possession, had been the loss of capacity of households in the region of Transvaal for agricultural subsistence through a confluence of state-imposed land dispossession and resettlement (Wolpe 1972, 427). Household subsistence needs had to be met through the labour migration of some of its (primarily male) members. As a result of these political-economic processes, the extended family has not only become the locus of social reproduction in the rural area but also of reproduction of migrant work force for the urban-based economy—which could thereby be paid lower wages (Wolpe 1972, 435). This political-economic development, buttressed by severe laws of racial segregation, has posed a challenge to

---

[8] The cloths used in Venda ancestor possession show remarkable similarity with those used in Karanga demonic possession (Werbner 1989, 61–108), from whom some aspects of ancestor possession have been adopted.

[9] The evidence of this state of social fragmentation derives from the Archives of the Berlin Mission Society.

[10] Sometime between the fieldwork of Stayt (1931) conducted in late 1920s and field researched of late 1950s by Blacking (1985).

the continuity of the extended family as it has—in the 'Venda' case—become stretched over primarily two loci: 'Venda' ('homeland') and *Tshikuani* (Johannesburg and Pretoria), with women, children and the infirm primarily residing in the former, able-bodied men in the latter.[11] It has also threatened former locally defined identities and allegiances as family members were forcefully relocated or were out-migrating to the cities to work for wages. The region of north-eastern Transvaal has been assigned the place of a periphery in the racially indexed political economy, centred in the Gauteng mining and industrial complexes and government buildings.

It is in this context that the household in 'Venda' has become a site of a new sort of 'social drama' (Turner 1981, 24) in which primarily rurally residing women have figured as mediums for channelling a sense of continuity between living descendants and their dead—now clearly genealogically defined persons *vis-à-vis* the possessed ego. As the extended family has contracted through the leaving of its members as labour migrants and through policies of forced resettlement in the 'homeland' in the present, its membership has been expanded in its temporal dimension towards the (unchangeable) past of genealogical continuity. Moreover, ancestor possession has become a new way of commemorating the dead, away from the official powers of the chiefs and the state (see also Blacking 1985). Women, primarily, have used ancestor possession to address predominantly issues of problematic fertility. However, confronting conditions of women's hazardous fertility has not simply constituted a matter of the private, domestic sphere located in the rural 'reserve'—a sphere somehow pre-existing symbolic actions engaging the ancestor spirits. On the contrary, 'domestic domain', significantly challenged by the wider processes of the racially indexed political economy of rural–urban labour migration and forced resettlement schemes, has been re-imagined through ancestor possession.

As the historical sources have attested, ancestor possession has not represented an essentialized part of the 'Venda' cultural repertoire of encompassing healing interventions which would be available as cultural

---

[11] Especially since 1980s, women have also preponderated as labour migrants to Johannesburg and Pretoria.

resources, prior to major social and political change impacting the 'Venda' region. Ancestor possession has emerged in the midst of transformations drawing Tshivenda-speakers of the north-eastern Transvaal ever more tightly into the political economy of South Africa. Streamlined possession phenomena have emerged as a way of coming to grasps with these processes, channelling a revaluation of the domestic domain, the significance of kinship and 'Venda' territorial autonomy. The primary symbols of possession—ancestor spirits—have been constituted as the exclusive agents of misfortune during the 1950s among predominantly rurally residing women experiencing fertility disorders. During the 1960s, a decade after such solidification of the ancestor possession 'cult', a small number of men in the capacity as labour migrants in *Tshikuani* have experienced symptoms identified as ancestor possession. The symptoms of affliction, as described in the previous section, manifested in the experiences of male migrants in *Tshikuani* along with dreams of 'leopard skins' and of 'returning to the mother's land as king'. In informal interviews, the men invariably stated that they wished to have their possession ritual expedited despite protests of their rurally residing female relatives as to their young age. They were eager to undergo initiation and stay in 'Venda' as practicing healers. Invariably, their wishes were thwarted. They had possession rituals undertaken on their behalf by the rurally residing mothers who mobilized their networks of female relatives, neighbours and friends, versed in possession undertakings; ancestor spirits descended into the men's bodies as matter of 'cure' which was sufficient at this point. The relatives of the young men—with the latter's reluctant consent—called the ancestor spirits into a pouch and hung it from a branch of a fig tree, pleading with them to allow the young men to return to *Tshikuani* and finish their work careers. Only after retirement from *Tshikuani* employment did the men become healers residing in the rural areas of 'Venda'—usually by the end of the 1980s.

Contemporary Transformations: Immobility in the Rural 'Home'

The period of the 1990s has seen a significant increase in the number of both men and women incapacitated by ancestor spirits at the workplace and in the life-world of urban employment in *Tshikuani*. This group of migrants has differed from the previous singular cases of ancestor possession afflicting migrants in the urban centres in two respects. Firstly, since the 1990s, migrants diagnosed as suffering from ancestor possession have dreaded the repercussions of this diagnosis, rather than encouraging it—as their predecessors had done. Secondly, the earlier generations of possessed migrants articulated willingness to return to the rural area of 'Venda', resettle therein and take up the profession of 'traditional' healer. Such aspirations of migrants possessed between 1960s and 1980s had been thwarted through negotiations of their kin with the ancestor spirits, resulting in the migrants' return to urban employment. In contrast, migrants encountering difficulties in *Tshikuani* since 1990s until the mid-2000s—complications diagnosed as ancestor possession—were compelled by this diagnosis to return to a 'rural' location within 'Venda' as a matter of avoiding an existence 'tethered between life and death'. In other words, what had constituted a negotiable 'ancestral wish' whereby the location of the possessed migrant could be determined according to the wishes of the living humans (the migrants' relatives) has subsequently become a strictly non-negotiable stipulation understood to be demanded by the ancestors. Territorial transfer from the urban centres of *Tshikuani* to one or another 'rural' location within 'Venda' has become a determining precondition for the re-gaining of health by the migrants afflicted in the urban centres of Gauteng.

This contemporary development of ancestor possession is reminiscent of analytical observations as to the continuing importance of rural 'homes' for migrants in South Africa (Posel 2003; James 2001). Analysts have pointed out that 'altruism' as well as 'self-regarding motivations' when rural households have provided 'insurance against unemployment' and place of 'long-term retirement' have constituted factors in urban migrant's continuing relations with rural 'homes' (Posel 2003, 3). James has suggested that homes in rural areas have constituted places 'to which children could be sent to learn traditionalist values…and the necessity of respecting one's elders' (James 2001, 103); they have therefore been maintained, often together with urban houses, as part of a strategy of locating family members according to their position in the kinship network and the life cycle (James 2001, 101). Rural homes have been associated with a positive moral dimension in contradistinction to 'the evil' of towns (James 2001, 101–102). Furthermore, as Lee (2011) has noted for migrants in Cape Town, visitations of the dead in dreams of their descendants have occasioned exhumations and reburials from the 'cold' town cemeteries to the 'warmth' of rural areas of Eastern Cape as part of 'reimagination of belonging' (Lee 2011, 240).

Such interpretations of continuing or renewed significance of rural areas as places of 'proper homes' for persons with migrant background at various stages of the life cycle have articulated an important feature of the 'new' South Africa. In this context, permanent settlement of families and migrants in places of work—that is primarily urban areas—enabled by the dismantling of Influx Controls in late 1980s and of apartheid itself in 1994 has not occurred, in contradistinction to the expectations of social analysts (Posel 2004, 277). 'Venda' ancestor possession raises the issues of motivations of urban–rural returns and the character of rural 'homes' anew as it offers a number of qualifications to the previous accounts as well as a new dynamic to the rural–urban dichotomy. Migrants are not only impelled by the ancestors to leave *Tshikuani* at the height of their work careers rather than at their consumption (for retirement, end of work contract, death), but further enact immobility within the rural area of the former 'homeland' as a way of remaining 'healthy' for the rest of their lives—as they have claimed in mid-2000s. Their *Tshikuani* home is abandoned in favour of building of a household explicitly in the 'rural' area, ideally expanding into several house structures. Enacting immobility in 'Venda' as a way to 'cure' is invariably stated as an option which was 'forced' upon the migrants through debilitating illness, rather than being a matter of positive choice. The interpretation of 'cultural revivalism' (James 2001, 103) more widely noted for South Africa is clearly relevant at some level but does not address the peculiarities of the experiences of the afflicted migrants nor of their relatives and wider communities.

Venda-*Tshikuani*: Past and Present Dichotomies

The historical background has been introduced as a way of showing that Venda ancestor possession, rather than constituting a static 'tradition', has recovered an orientation to clearly defined genealogical 'pastness', as substantial transformations threatened the constitution and continuity of the extended family and undermined economic subsistence of households. A turn to an unchangeable past defined in terms of the dead generations has been a way of grounding the changing present, as well as expanding its dimensions (see also James 1999, 168). Moreover, soon after its inception, common relations to own ancestor spirits have emerged as a way of relating both rural women and male labour migrants in the same set of activities—possession rituals and healing practices, associated with the household within 'Venda' territory. A sense of genealogical continuity identified as the cause of afflictions once broken has served to at least periodically patch over the geographical distances, entrenched in the regime of labour migration, between the rural women and the migrant men. However, different sorts of limits were set upon the migrants possessed by ancestor spirits in *Tshikuani*. While during the apartheid period, social consensus overrode the migrants' wishes to resettle in 'Venda', towards apartheid's end it has equally powerfully overturned the migrants' wishes to stay in *Tshikuani*. Articulated in different terms, while ancestor spirits could be negotiated with in the period of apartheid, they could no longer be placated during its aftermath as to the location of the descendant. In this section of the chapter I will try to elucidate how contemporary experiences of affliction and healing occurring in two sets of places—'place of white ways' and the 'homeland'—articulate with an earlier, 'dualist debate'.

Ancestor spirit possession of Tshivenda-speaking labour migrants maps unequivocally onto the 'dualist debate' which has powered significant arguments within social anthropology of southern Africa. This discussion has several facets which all in various ways deal with the problem of how men and women have come to inhabit a dissonant landscape of rural–urban labour migration. The divisions of this landscape were buttressed by severe laws dictating where categories of persons could be located geographically between the 'city' and the 'homeland'.[12] According to different authors, number of strategies has emerged as people entrenched in the divisions of the political economy, which simultaneously demanded their compliance as movable, cheap labour force, have come to inscribe such contradictory social experiences with cultural significance.

Jean and John Comaroff (1987) have argued that the Tshidi-Barolong dealt with their situation of encompassment as an exploited labour force and a people marginalized from centres of power through the creation of a set of 'poetic' contrasts. These hinged on the opposition between *sekgoa* and *Setswana*, 'things of whites' and 'Tswana tradition', respectively (Comaroff and Comaroff 1987, 195). Within these oppositions, another pair of contrasts was nested: that between 'work for whites' (*bereka*) and work as creative activity of self-enhancement (*tiro*). Through these, the Tshidi-Barolong have come to reflect on their experience of incorporation into the racially indexed political economy, firmly identifying the powers which engendered the former's subordinate status in the new set of forces of colonial and apartheid South Africa—the institutions of 'white'-dominated government and economy. Furthermore, the authors demonstrate that this set of contrasts has not amounted to a simple commentary of the Tshidi-Barolong on their historical circumstances. It simultaneously constituted the means for creating own history. The authors illustrate this process with the case of a migrant who chose to opt out of labour migration, carving own subsistence from the dry soil of the 'homeland' of Bophuthatswana: explicitly as a way of 'working for oneself' (*tiro*) in contradistinction to the alienating 'work for whites' for money (*madi*).[13] Jean and John Comaroff have thus articulated a dualist view of how a people encompassed within the racially indexed political economy of South Africa sought 'to remake a recognisable world beyond the predators' grasp' (Comaroff and Comaroff 1987, 201).

---

[12]This state of segregation, not only between 'whites' and 'blacks' but also within the African population itself—entrenching divisions of urban and rural folk, was brought about by a set of laws, particularly the Pass Laws and Influx Control of 1950s and 1960s and the Bantu Homelands Citizenship Act, 1970.

[13]The significance of the term 'money' which also denoted 'blood' was consciously used by the migrants to reflect on their experience as being drained of vital substance under the exploitative and dangerous conditions of work (Comaroff and Comaroff 1987, 200).

Working with Basotho labour migrants in South Africa towards the end of apartheid, Coplan (1994), on the other hand, has rejected the 'dual-worlds thesis' propounded by Jean and John Comaroff (1987) in the main part of their work. It was not their incorporation into the South African political economy which the Basotho migrants feared or tried to prevent during the apartheid era and towards its dissipation. They recognized Lesotho and South Africa as being part of a 'single social world' (Coplan 1994, 147). Nevertheless, mindful of the gross inequalities which they as 'foreigners' in South Africa have faced, Basotho labour migrants have deployed performative genres of 'auriture' ('word-song') as a way of creating 'an integrated, positive, cross-border self-concept [that] has been maintained in the face of systematic displacement, fragmentation, and dehumanization' (Coplan 1994, 144). Nevertheless, Coplan also maintains that Lesotho has constituted a location towards which the migrants have projected an idealized set of categories associated with Sesotho, a transforming tradition with signs, however, firmly oriented towards the pre-colonial past. Through the values and practices of Sesotho, migrants have been able to maintain a distinct self-concept and identity.

Although Coplan insists on the divergence of his account relating to Basotho migrants' identity from the dualist thesis of Jean and John Comaroff, he makes a somewhat similar finding. This relates to the importance of the 'country' (Lesotho)—if not as a separate economic world which would underpin resistance of the Basotho to the political economy of South Africa, then definitely as a distinct geographical locus regarded as a repository of proper Sesotho cultural values. A degree of separation, even if on the level of moral ideals rather than of mode of subsistence, is operative in Basotho migrants' self-concepts enfolded through the idealized world of cultural values and practices projected onto the geographical locale of Lesotho. The conclusion which can be drawn from these two cases, relating to the Tshidi-Barolong and the Basotho migrants, is the following: a geographical area associated with 'own culture' and with a degree of autonomy from the urban centres in South Africa has played an important part in migrants' attempts to (re)create cultural identity amidst the displacements engendered in the coercive regime of labour migration under apartheid.

The 'dualist debate' remains relevant also for the case of ancestor spirit possession afflicting 'Venda' labour migrants in post-apartheid South Africa. For the affliction associated with ancestor spirit possession for which the only remedy is resettlement to the territory of the former 'homeland' of 'Venda', does affirm the rural area as a repository of values in contradistinction to the urban centres. In this respect, the figure of the itinerant 'Venda' migrant becoming immobile within the 'homeland' as a matter of cure seems reminiscent of the case of the Tshidi-Barolong and Basotho migrants who also endow the 'country' with positive valuations in contradistinction to the urban centres. However, that in the 'Venda' case these values are *therapeutic* in character, rather than constituting moral notions, is of some note. Bodily symptoms and therapeutic interventions are not easily interpreted when it comes to clear-cut definitions as to their significance. Nevertheless, they raise a number of issues which are pertinent to the themes which have been sketched above under the heading of the 'dualist debate'.

Although it is in the urban vortex, *Tshikuani*, where migrants fall 'ill', this misfortunate condition is not articulated as somehow arising out of the 'evil' character of the cities themselves (as indicated by James 2001, 8), which would have a corrupting influence on the migrants. The symptoms of *Tshikuani* illness themselves suggest that the migrants may be traumatized by labour migration experience and expressing their discontent with the urban workplace, if we view the complex of the 'denying body' in western psychiatric terms. However, that would be taking symptoms out of their proper cultural context and denying their cultural and personal significance. For one, in the life histories of the migrants the 'city', *Tshikuani*, does not figure negatively as inherently an afflicting environment. Moreover, all of the migrants, both men and women, with whom I worked expressed pride as to their ability to keep the '*jobo*' in *Tshikuani* and thus become respectable breadwinners for their families. In another idiom, *Tshikuani* is also denoted as *tsheledeni* ('place of money'), that is a source of much desired resources and income. Rather than being incapacitated to enact urban employment, *Tshikuani* illness as it is interpreted through the idiom of ancestor spirit possession, affirms the migrants' capacity for urban work: for as spirits have seen fit to usurp this capability, the migrants themselves are deemed worthy of urban employment and are not 'defunct' men and women (cf. Boddy 1989, 255).

The 'Venda' case of ancestor spirit possession contributes further material to the dualist debate confirming that the divisions of the 'city' and the 'homeland' have been endowed with renewed significance since the dissolution of apartheid in early 1990s. Mutually enmeshed in a single therapeutic landscape through the symbolic mediations associated with ancestor spirits, the 'city' and the 'country' have been assigned opposing values—as health-detrimental and health-promoting, respectively. However, what are risks and promoters of health are not places as essentialist milieus, in the sense that a 'miasma' of the city would produce illness in its inhabitants (Sontag 1978) while the landscape of the countryside would engender health for its occupants. The degree of health-promotion of specific places inheres rather in their territorial quality, as markers of geographical loci that have been shaped historically through colonialism and apartheid: the urban vortex of Gauteng (*Tshikuani*) and the 'homeland' of 'Venda'. It is via the physical location of the possessed migrants within this historically rich geography that 'illness' and 'health' have been articulated. In the 'Venda' cultural context, medical concerns have thus emerged as vehicles of a dysphoric geography, as the affective tracings of apartheid's geographical divisions. These divisions have been recast in present practices of ancestor possession whereby the territory of the former 'Venda homeland' has been revalued: not as an essentialized repository of cultural values but as the site of dramatic encounters between humans and their ancestors through which proper directions for sociocultural change have been negotiated.

Conclusion

Therapeutic immobility in 'Venda' mediated by ancestor possession for Tshivenda-speaking labour migrants falling 'ill' in *Tshikuani* has not constituted a form of immobility easily predicted by macro-level social science studies associated with the 'mobilities turn'. Moreover, the resettlement which it has entailed *Tshikuani*-'Venda' transfer of afflicted migrants in order to effect 'cure' has not amounted to a simple socio-spatial move towards the embrace by 'home' community of persons alienated in the capitalist workplace. Much 'work of culture' (Obeyesekere 1990) has gone into contemporary therapeutic immobility in a highly mobile 'new' South Africa, condensing historical experiences of a divided (colonial and apartheid) world for promoting cultural creativity in the present.

Immobility, associated with the radical curbing of movement in the 'Venda' cultural context, is indeed the result of a protracted process of coming to grasps with the ancestral 'gift', *mpho*, which the migrants have tried to 'dodge' for many years. However, immobility—which has a crucial geographical component—is not simultaneously associated with a sense of 'being stuck' or frustration as to being unable to move according to one's wishes. It is a positively valued condition of possessing the 'gift'—bestowing clear vision and healing capacities by ancestors onto the living kin they have chosen; a capacity of clairvoyance that can only derive from supra-human forces. The chapter has demonstrated that therapeutically mediated immobility is the result of negotiations between humans and their ancestor spirits and it thus constitutes an active state. Moreover, immobility must be constantly (re)negotiated as properly enacted. As ancestral mediums, persons cannot move of own accord. They also cannot 'sit still'. Ideally, an ancestral medium is constantly busy with divination and healing activities at least partially visible to public scrutiny.

References

Blacking, John. 1985. "The Context of Venda Possession Music: Reflections on the Effectiveness of Symbols." *Yearbook for Traditional Music* 17: 64–87.

Boddy, Janice. 1989. *Wombs and Alien Spirits: Women, Men, and the Zar Cult in Northern Sudan*. Madison: University of Wisconsin Press.

Comaroff, Jean, and John Comaroff. 1987. "The Madman and the Migrant: Work and Labour in the Historical Consciousness of a South African People". *American Ethnologist* 14 (2): 191–209.

Coplan, David. 1994. *In the Time of Cannibals: The Word Music of South Africa's Basotho Migrants*. Chicago: Chicago University Press.
Good, Byron, and Arthur Kleinman. 1985. *Culture and Depression: Studies in the Anthropology and Cross-Cultural Psychiatry of Affect and Disorder*. Berkeley: University of California Press.
James, Deborah. 1999. *Songs of the Women Migrants: Performance and Identity in South Africa*. Edinburgh: Edinburgh University Press.
James, Deborah. 2001. "Land for the Landless: Conflicting Images of Rural and Urban in South Africa's Land Reform Programme." *Journal of Contemporary African Studies* 19 (1): 93–109.
Khan, Nichola. 2013. "A Moving Heart: Querying a Singular Problem of 'Immobility' in Afghan Migration to the UK." *Medical Anthropology: Cross-Cultural Studies in Health and Illness* 32 (6): 518–534.
Khan, Nichola. 2016. "Immobility." In *Keywords of Mobility: Critical Engagements*, edited by Noel B. Salazar and Kiram Jayaram, 93–112. New York: Berghahn Books.
Lee, Rebekah. 2011. "Death 'on the Move': Funerals, Entrepreneurship and the Rural-Urban Nexus in South Africa." *Africa* 81 (2): 226–247.
Low, Setha. 1994. "Embodied Metaphors: Nerves as Lived Experience." In *Embodiment and Experience*, 139–162. Cambridge: Cambridge University Press.
Obeyesekere, Gananath. 1990. *The Work of Culture: Symbolic Transformation in Psychoanalysis and Anthropology*. Chicago: University of Chicago Press.
Ong, Aihwa. 1988. "The Production of Possession: Spirits and the Multinational Corporation in Malaysia." *American Ethnologist* 15 (1): 28–42.
Posel, Deborah. 2003. "What Has Been Happening to Internal Labour Migration in South Africa, 1993–1999?" *The South African Journal of Economics* 71 (3): 455–479.
Posel, Deborah. 2004. "Have Migration Patterns in Post-apartheid South Africa Changed?" *The Journal of Interdisciplinary Economics* 15: 277–292.
Scheper-Hughes, Nancy. 1993. *Death Without Weeping: The Violence of Everyday Life in Brazil*. Berkeley: University of California Press.
Sharp, Lesley. 1994. *The Possessed and the Dispossessed*. Berkeley: University of California Press.
Sontag, Susan. 1978. *Illness as Metaphor*. New York: Farrar, Straus and Giroux.
Stayt, Hugh. 1931. *The Bavenda*. London: International Institute of African Languages & Cultures.
Taussig, Michael. 1980. "Reification and the Consciousness of the Patient." *Social Science and Medicine* 14 (1): 3–13.

Turner, Victor. 1981. *Drums of Affliction. A Study of Religious Processes Among the Ndembu of Zambia.* Ithaca: Cornell University Press.

Werbner, Richard. 1989. *Ritual Passage, Sacred Journey: The Process and Organization of Religious Movement.* Washington, DC: Smithsonian Institution Press.

Wolpe, Harold. 1972. "Capitalism and Cheap Labour-Power in South Africa: From Segregation to Apartheid." *Economy and Society* 1 (4): 425–456.

7

Liminality and the SCI Body: How Medicine Reproduces Stuckedness

Colleen McMillan

This chapter explores these questions based upon the author's clinical and academic practice with individuals living with SCI who experience medical regimes and protocols imposed upon them through multiple avenues; healthcare professionals, clinical practices and institutional policies. Thomassen's (2015) typology of liminal experiences defined *by the subject* (individual, group or society) or the *temporal dimension* (moment, period and epoch or lifespan) will serve as a reference point in which to better understand such questions (Horvath et al. 2015, 49). The typology will be used to explain how larger societal narratives of normality perpetuate the concept of immobility to this population whose mobility has been altered. Equally important, this chapter offers a place for the voices of those who are immobilized to offer insight into their lived experience of liminality. In this way, this chapter represents a space for such voices to become unstuck, where the individual's resilience becomes visible and change can occur through awareness. Viewing the SCI body as a

---

C. McMillan (✉)
Renison University College, University of Waterloo, Waterloo, ON, Canada
e-mail: c7mcmill@uwaterloo.ca

partner, rather than a problem, starts the critical conversation on how the paradigm of healing can shift to a collaborative, person-centred process.

Introduction

I was told by a SCI patient that the sensation of experiencing a spinal cord injury is not unlike "*the tires being ripped off a car. The body was still there but what use it is if it can't move?*" (C. McLean, personal communication, May 17, 2016). While it is commonly assumed that acute trauma to the cervical spinal cord inevitably results in total body or complete paralysis, this is not always the case. The location of the trauma to the spinal cord predicts the degree of paralysis or immobility, and whether future movement will be intermittent or complete. However, both physicians and patients alike will say that nothing is absolute in the context of SCI trauma, because the body enters an evolving liminal space held together by multiple phases of stuckedness and unstuckedness emotionally bookmarked between hope and disappointment.

In cases where the cervical spinal cord is not completely severed, or if severance is incomplete, then the individual is diagnosed with partial paralysis. Although this term evokes more optimism for recovery or rehabilitation, it brings with it a unique set of challenges to be unpacked. Unlike complete paralysis, with partial paralysis the individual may temporarily regain some functioning below the site of injury, only to later lose it. Partial paralysis removes the ability to feel or move the legs or feet, walk, perform bowel and bladder acts and perform sexually. Partial paralysis suspends the body into a perpetual state of negotiation between corporeal boundaries that are fluid and stuck at the same time.

Perhaps no greater single act of injury renders the body so utterly dependent upon the beneficence of others, and a lifelong dependency upon the medical system. A spinal cord injury also ushers in a lifelong state of liminality; an anthropological concept that refers to "breaking boundaries and threshold experiences" (Horvath et al. 2015, 7). With injury, the body as a single, contained entity no longer exists. Instead, various and multiple anatomical appendences and organs either shift organically or are moved mechanically to test new limits of functioning.

The body enters a perpetual state of testing thresholds looking for new ways of adapting to the reality of disability.

Thomassen's (2015) concept of the *individual moment*, described as a sudden event occurring in one's life, is particularly effective in conceptualizing how the body is forever changed upon injury (Horvath et al. 2015, 49). To appreciate how his concept becomes embodied into a new normalcy and appreciate to what degree the medical model is imposed into the daily management of the body, it is important to understand the number and complexity of anatomical systems that are affected by a SCI injury.

To begin with, both the central and peripheral systems are compromised in addition to the skeletal, urinary tract, digestive, respiratory and autonomic systems. Skin and sexual functioning, including that of fertility, is also impacted. The transitionality of symptoms exacerbated by the likelihood of complications (Sezer et al. 2015) positions the individual with SCI towards a trajectory of mental health issues unlike any other health issues. Several studies have found that depression and anxiety can emerge up to 15 years post-rehabilitation. Prevalence of depression falls between 20 and 30%, which is three times higher than the general population (Krueger et al. 2013). Similarly, prevalence rates of anxiety among those with SCI range from 10 to 35% (Banerjea et al. 2009). In this context, it is not surprising the rate of suicide post injury among this population is three time higher than that of the general population (Cao et al. 2014; Lim et al. 2017).

An evolving army of medical specialists becomes synonymous to living with a spinal cord injury, and offers a glimpse into the complicated relationship between medicine and disability via immobility. What develops is a medicinal and pharmaceutical interplay where immobile appendages and organs are urged and probed to move, assume different shapes, and perform new functions, only to inevitably reach plateaus where stuckedness emerges or inevitably returns. These *individual moments* eventually become *temporal* segments of time alternately moving towards or away from the societal defined boundary of normalcy.

This chapter explores how different facets of medicine specific to the care of the body traumatized by a spinal cord injury perpetuate a state of immobility and emphasize the stuckness of individual parts. I will explore how medical systems suspend not just the SCI body per se, but different parts of the body into a state of liminality with concurrent cycles of stuckness and unstuckness becoming the new norm. In this scenario, a host of medical mechanics work on body parts with the goal of repairing, but never really fixing. In this way, it is impossible for the body as a unit to feel completely unstuck, because, returning to the metaphor of the car, parts are addressed in isolation rather than a whole.

Starting at the point of trauma, the chapter will explore how medical systems construct the landscape of liminality for the individual with SCI. Drawing upon existing literature and first-hand clinical experience working with SCI patients, this chapter examines the relationship between medicine and immobility of the body with SCI. Space will be created to make visible the voices of those who live with complete or partial spinal cord injury. Voices that permanently exist within the confines of immobility can contribute the authenticity of existence, and are critical for disrupting the homogenizing nature of the medical model that governs the SCI body.

Liminal Passages Experienced by the SCI Body

What happens when physical mobility stops because of trauma? How does one negotiate the millisecond shift from possession of one's able body to relinquishing the body to others? And how does ownership leave the body to be appropriated by a small army of interventionist medical systems? The above questions will be placed into the sequence of actions that occur once a body sustains the acute injury of spinal cord severance. Tracking the body as it moves through the different types of liminal experiences described by Thomassen (2015) will illuminate and contextualize the multiple and sequential spaces of mobility loss juxtaposed by the inhabitation of medical personal and protocols starting at the scene of trauma (Horvath et al. 2015).

The journey begins in the emergency room (ER), a space that functions as the visual and physical marker of the medical milieu for a SCI injury. Upon entry into the ER, trauma specialists immediately take possession of the body to examine and assess critical signs such as breathing, heart rate and blood pressure. If a spinal cord injury is suspected, a neurologist will be consulted to evaluate the degree of damage existing below the site of trauma by measuring motor function of ten muscle groups in the elbow, fingers, hips, knees, ankles and toes. Sensation is also measured across 28 sensory zones. Due to the fluctuation of scores, this test can be done up to several times a day, mapping out shifting zones of mobility/immobility. Nerve cells and groups of nerves called tracts will be monitored by a *physiatrist* for changes as a result of fluctuating inflammation and compression, risk conditions for infection and cell death. If breathing is severely compromised, as in 80% of cases (Bonner and Smith 2013), a respirologist will conduct an emergency tracheotomy to establish an airway. A series of CT and MRI scans will be ordered and read by a radiologist to assess for soft tissue and ligament injury as well as bone damage. These tests will indicate whether surgery is imminent by a neurosurgeon within the first 24 hours or after a few days. If bone shards are intruding into the spinal cord or the spine is not stable, an orthopaedic surgeon will be required to remove bone from the hip to stabilize the spinal cord, augmented with metal plates, screws and rods. The disruption of normal blood flow caused by the injury requires a pulmonologist to monitor for the prevention of deep vein thrombosis or blood clotting in the lower extremities. A urologist will be present to determine the need for catheterization which will allow for renal clearance, but also to avoid bladder over distension. This array and intensity of medical intervention dictates the first 48 hours in an effort to understand how the once mobile body is responding to the trauma of acute injury and the degree to which the body becomes tentatively referred to as either complete, incomplete or labelled with "a C6 or C7". This medical label now assigns the individual to a *group* and *period,* representing the second stage in Thomassen's (2015) typology (Horvath et al. 2015).

With the newly assigned label the body is transferred to another stage of medical intervention moving from the ER to the Intensive Care Unit (ICU) of the hospital. There a new team of physicians and specialists wait to assume ownership. At this point in time there is growing speculation regarding the degree to which the body is determined to be stuck or unstuck, a critical pathway in mapping a direction for future care.

The body resides in this specialized system for approximately 11 days where the focus is on stabilizing the spine, monitoring and responding to infections, managing neuropathic pain and assessing for changes in sensation, if any. Once stable, the body is then transferred to the medical floor where it waits for discharge to a specialized rehabilitation hospital.

A longer period is spent in a specialized rehabilitation hospital where the body continues to heal, adjust, and experiences some mobility movement forward, accompanied by sporadic and unpredictable retreats. At this stage, it is generally accepted by the different medical specialists that the pre-accident body is gone, and attention now shifts to categorizing body parts and organs to those that are stuck as compared to those that are not. The car metaphor is again useful here; in a broken car, the stuck parts are discarded and replaced with new ones. The unstuck ones are ignored. Similarly, in the SCI body the stuck parts gain new visibility for their brokenness, and unstuck parts are ignored. A commonality shared between the two is the assignment of value; a higher value and worth is associated with a car and body that is mobile, very little when stuck.

Finally, after many months, the body is discharged back to the community. This last transition marks the end of the expert interventionist period and can be likened to Horvath's dimension of the *epoch* or "lifespan duration", where the body "stands outside society by assignment" (Horvath et al. 2015, 49). The social expectations linking the ritual of independency to adulthood no longer fit and must be re-aligned. Another team of allied health staff now routinely enter into the most personal and intimate spheres of daily life: the home. With mechanical precision, they perform bowel and bladder routines, clean ostomy bags, assist with washing and eating, monitor the skin for the emergence of weeping ulcers and administer a dazzling array of pharmaceuticals designed to address chronic and neuropathic pain, uncontrollable body spasms and respiratory infections. The body defaults to a state of dependency orchestrated by a host of rotating strangers that record which parts are stuck or are unresponsive through daily surveillance.

Despite moving everywhere across the spectrum of medical expertise and levels of care (tertiary, rehabilitative, primary), the body actually "goes nowhere" (Hage 2009). In other words, the immobile body enters a liminal state. The earlier frenzy of medical intervention starting in

ER is over as there is nothing else to examine, replace or repair. The body is relegated to the hinterland of stillness. This new state is soon occupied by a mind that is perpetually awake and filled with emotion, searching for new constructions of meaning to justify existence. For the most part, the body is largely ignored by the medical model as it is no longer active, moving or responsive but rather a passive entity bereft of agency. Extending across the liminal dimensions of *moment, period* and *lifespan* defined by Thomassen (2015), the SCI body passes through these different dimensions in a matter of months; from the acute individual moment of injury to life cycle periods reaching across the epoch or lifespan (Horvath et al. 2015).

Medical Relationship to the Post Mobility Body

In the lexicon of medical language, the individual diagnosed with a SCI becomes a body known as disabled. The medical focus of care shifts from supporting the normal body to surveillance of the disabled body. In this scenario, the body is approached as the semblance of parts versus functioning of the whole. This perspective mirrors the positivist paradigm that houses the medical model, guided by beliefs that are embedded into clinical practice: disability is a problem that resides in the individual, it is a defect of a bodily function, and is "inherently abnormal and pathological" (Olkin 1999, 26). Goals for the post mobility body reflect the passivity in which the body is now seen. The SCI body is expected to adjust to the new disability situated in an unfamiliar environment, and is viewed as the passive receivership of medical services. The presumption exists that the individual will quietly acquiesce to the patient role because to resist is to risk being labelled as difficult or non-compliant.

Thomas and Woods (2003, 15) refer to the medical model of disability as the "personal tragedy" model because it defines disability in a fundamentally negative way. The medical model links a disability diagnosis to the physical body (Fisher and Goodley 2007), and perceives immobility as a "pitiable condition, or a personal tragedy for the individual, something to be prevented and if possible cured" (Carlson 2010, 5). In the case of the SCI body, cure is impossible, and the body is forever divorced from the norm. Coupled with medical terminology that describe the immobile as "invalid, handicapped and spastic" (Creamer 2009, 22), the SCI body can never be comparable with able-bodied counterparts. Johnstone (2012, 16) captures the binary construction of this model in starker terms with able-bodies seen as "superior" at one end of the spectrum as compared to disabled bodies as inferior at the other end.

Unfortunately, the statistics related to SCI accidents confirm that permanent loss is a new reality. A 2017 study identified that 53% of individuals are left as tetraplegic (partial or total paralysis of the arms, legs and torso) and 42% become paraplegic (partial or paralysis of the legs) (Sweis and Biller 2017). While this understandably is experienced as a loss of independence for the patient, it can also be experienced as a loss to the attending physician. Thomas and Woods (2003, 15) explain that physicians are trained to treat people as "problems to be fixed". In the case of the SCI person, however, there is *nothing to fix*. Regardless of the amount of knowledge or training, a physician can never make a person with SCI walk again, jump in the air, toilet themselves, prepare a meal or hug a child. This situation becomes a quandary forced upon both parties; the physician who is at a loss of what to do with stuck body parts, and the patient who has been socialized to believe in the miracle of medicine and the archetype of the good patient (Campbell et al. 2015). Without something to fix or make move again, the physician retreats and becomes uncertain how to manage the stuck body. A new tension is borne. Ironically the medical model that continues to dominate clinical practice for SCI patients may not be working well for either physician or patient.

In a study that explored how physicians feel about working with individuals with spinal cord injury several themes emerged as to the different forms this tension assumes within the physician–patient relationship (McMillan et al. 2016). One way this tension takes form is through indiscriminately downloading dominant expectations onto the individual SCI patient, by assuming the stuck body can somehow fit able-bodied office apparatus. The study surveyed family physicians tasked with the care of individuals with SCI and found that basic examination equipment did not exist in their offices to care for this population.

Despite almost all the physicians in this study describing their medical practices as being physically accessible, handicapped washrooms, automatic doors, ground floor locations, inaccessible parking and adequate door entrances were absent. For the small number of physicians who had purchased portable tables that would allow for a physical examination, it could not be remembered where the table was stored, as reflected by this physician's statement:

> You know I'm not sure. I'm not sure. I think we might have some [equipment] that is portable, but again, because they're really never used, I'm not sure whether we do or not or even where it is located. (McMillan et al. 2016, 467)

Similarly, the physicians who reported their practice had a height adjustable examination table, in the absence of a transfer mechanism (e.g. lift), which none of the physicians reported having, patients with more severe mobility impairments were unable to get on the examination table preventing physical examination of the lower extremities:

> they all need step stools to get on to, we don't have any beds that lower down to a lower level to get on, so it usually means we have a couple of nurses who have to assist someone to get on the examining bed. (McMillan et al. 2016, 468)

These findings concur with Kasser and Lytle (2005, 110) who note that the medical model regards the person with a disability as the "one who needs to change or be fixed, not the conditions that might be contributing to the person's dis-ability". They put forth the proposition that medicine "essentially disregards the environment that might intensify or adversely affect a person's functional abilities" (Kasser and Lytle 2005, 11). Thomassen's (2015) dimension of *society* as part of the liminality typology offers another explanation. The provision of standardized medical equipment designed for the able-bodied patient highlights the "reproduction of social and political structure" within medicine (Horvath et al. 2015, 49). Physician offices are arranged around the dominant values of time, autonomy and independence and operationalize these values into clinical practices. This sentiment is captured by a physician in primary care in regard to the office environment, and not having an accessible examination table or weight scales.

> The procedure room with the high/low table is actually quite far away, so if we need to access it, convenience can be a bit of an issue sometimes to get in there. (McMillan et al. 2016, 467)

Embedded in this statement is the notion of convenience, but for whom does the lack of examination equipment pose an inconvenience? The notion of provider convenience can also be extended to how time is perceived.

> When you're dealing with someone with spinal cord injury just the transfer from their chair to the examining table can take sometimes, 15 to 20 minutes, so when you're running a practice where you need to see lots of people that can take up a lot of your time. (McMillan et al. 2016, 468)

The notion of money emerged as another theme as highlighted by the physician statement:

> ...a patient that takes you five times longer is not the patient you're encouraged to look after better. Um, that's got everything with to do the way the system has remunerated you. (Joseph et al. 2018, 70)

Tensions associated with the concepts of time, convenience and money converge to infer taking care of the SCI stuck or disabled body is problematic. As a result, spaces of liminality are routinely replicated with the outcome of disempowering the SCI body. By translating these tensions into expectations of adjustment onto the individual patient, these examples highlight how the medical model supports the notion that the immobile body needs to adjust to the existing environment, rather than the reverse, and, in this way, reproduces and makes visible the stuckness of the paralysed limbs and appendences.

The Western value placed on the commodities of time, convenience and money intersect to being incompatible to the patient whose body parts are stuck, don't move and cannot be configured to conventional medical apparatus. Disabled bodies are frequently spoke of as being a poor fit for mainstream services that cater to the able body, preventing some of the most basic preventative health tasks to occur, such as being weighted or pelvic exams being done, as noted by this one individual in a wheelchair,

> I have frequent visits to a doctor's office, neurologist, physiatrist, urologist, you name it, primary care, you name it. I see quite a few doctors and not one of them has a way of weighing me. (Disability Rights and Education Fund, n.d.)

In situations of reproductive and sexual health for women with SCI, it is often reduced to what is most convenient to the attending physician, who may make unilateral decisions without attaining consent, as noted in this statement:

> I've had patients that… um …are so spastic that their legs can't be open to have a pelvic exam…We were in the operating room and I thought I will have to do an abdominal [rather than vaginal] hysterectomy cause her legs couldn't be opened. (Joseph et al. 2018, 67)

Medical environments are structured around servicing the able bodied and are shaped to varying degrees by the hegemonic triad: time, convenience and money. A body with stuck parts defies such conventions and end up bearing the burden of responsibility to receive the most basic of care. That said, the historical structure in which health care is delivered is problematic because of how it leaves the physician and the patient with unrealistic expectations of the other, setting up a tension between individuals rather than systems.

The Reproduction of the Medical Model

In a book that advocates for a holistic approach by medical practitioners towards suffering and healing, Cassell (2011, 12) proposes that one first starts with a definition of a person:

> a person is an embodied, purposeful, thinking, feeling, emotional, reflective, relational very complex human individual of a certain personality and temperament, existing through time in a narrative sense, whose life in all spheres points both outward and inward *and who does things*. A person is always in action and never quiescent.

If Cassell's definition holds true as a normative viewpoint of personhood for clinicians, then such an endorsement exemplifies the problematic underpinning of medical curricula and illustrates how reproduction of the medical model occurs; how care is conceptualized and delivered to those individuals whose bodies deviate from the anatomically perfect and mobile body. The descriptive definition of a person being "always in action" and "never quiescent" suggests that the natural movement of the body should be ever ceaseless and fluid, as opposed to still and stuck. In an article published in the Lancet, "*Disability and the training of health professionals*", Shakespeare et al. state "curricula medicalise disability [and] fail to take a holistic view of health" (Shakespeare et al. 2009, 1815). Likewise, Joseph et al. (2018) argue that physicians might approach patients with disabilities differently if critical disability studies were part of their medical education. The question then becomes why medical schools remain reluctant to include course(s) on critical disability. While it was previously noted how the medical model commodifies the concepts of time, convenience and money, Thomas and Woods (2003) offer another explanation: power. The medical model of disability assigns tremendous power to physicians who diagnose people based upon criteria developed from constructions of normalcy (Thomas and Woods 2003, 15). Medicine has historically been synonymous with power (Canter 2001) and medical schools are hesitant to challenge prejudice (Evans 2004) as evidenced by the quote:

There's a lot of, sort of defensiveness, I think. We take pride in the fact that we're a self- regulated profession and believe that you know, everything we do is the best we can do. So, it's sometimes more difficult to digest, um, that maybe we're not doing that, that we could be doing better for a population of patients that we as a, as a community of physicians may be marginalizing without even realizing it. (Joseph et al. 2018, 73)

While a paradigm shift towards a social disability model has been heralded by some, this quote suggests a degree of defensiveness remains entrenched within medical teaching institutions to acknowledge power, let alone relinquish it. Despite language that appears holistic and suggestive of inclusivity through the use of such terms as "whole person care" (Hutchinson 2017), "person centered care" (Stewart et al. 2013) and "participatory medicine" (Rogers et al. 2013), the voices of those who are on the receiving end raise doubt at how successful such initiatives actually are. It appears incongruency exists between the progressive principles advocated for and the reality of medical practice that continues to focus on individual pathology.

In a small pilot study that explored how traditional screening tools assessed individuals with SCI of their healthcare experiences, one of the key themes to emerge was that questions in the questionnaires were too focused on pathology and the physical aspects of their disability (McMillan et al. 2018). Of the 17 individuals interviewed, 14 stated the issues that were important to them were either missed or left out by the questionnaires. Instead they expressed a level of frustration by being asked only about body parts (bowel, bladder, lungs) or problems (mobility challenges) that focused on or emphasized the stuckness of body parts. A follow up survey with the same individuals revealed stories of positive adaptation or anti-narratives were left untouched. As an example, one participant stated,

> I like the person I've become and I realize that my spiritual growth and I think it is almost a kid of a minor evolution in the human spirit to deal with challenges and to not get held back by these challenges. P[1]

Similarly, another individual shared:

> I've become a much more independent person, I'm a much more confident person. I was always a social person, but I found purpose in life, I think. P[7]

Furthermore, for this group of individuals immobility was not a constraint in the pursuit of a social life. The ability to assimilate the physical outcomes of the SCI injury into a reformulated sense of self is noted by P[2] who, despite being confined to a wheelchair with noticeably visible muscle atrophy, said;

> like a lot of people think, oh, you ended up in a wheelchair and you got to stay inside. Like, I am out every day, every weekend, and I'm at the club once a month, and sure you're going to stand out, but like goes on.

Lastly, the language used by this group of participants used to self-report their state of health defies the constructions found in the literature. Phrases such as "handi-capped, abnormal, vegetable and spastic and confined to a wheelchair" have been used to describe those who are restricted by immobile body parts and reflect stigma and stereotypes by healthcare providers (Smith 2009; FitzGerald and Hurst 2017). In contrast, individuals who were either paraplegic or tetraplegic shared descriptions of their general state of health in ways that spoke of reconstituting the stuck parts into a re-envisioned sense of self. Ten of the 17 individuals described their new post-SCI body as "really good", "normal" or "exceptional". In fact, one individual extended his description to suggest that the injury was not traumatic at all but instead ushered in unexpected outcomes.

P [9] I've never had a trauma, sure it was a physical trauma but I love what I was doing and I knew there was a high risk and that why I did it. What has happened is that my brain has actually taken over so my abstract thinking has really taken over squatters rights on the sensory part so I'm in awe of what's going on.

[I] so, you didn't experience [the SCI] as traumatic? Or that the body parts that you can no longer use as traumatic?

P [9] absolutely not.

Traditional screening instruments that fail to hold space to collect anti-narratives and detail how stuck body parts are positively reconceptualized, risks the likelihood that universal and pathologizing constructions of the SCI body are reproduced.

Spaces of Reproduction and Resistance

Spaces where aspects of the medical model reproduce or further entrench the stuckness of the SCI body are multiple and generally appear not to be shifting. Such spaces assume multiple shapes ranging from the practical (not having adequate office equipment to do basic preventative examinations) to outdated pedagogy (curriculum that purports only one model of disability exists that is disempowering to marginalized individuals) and finally, the current use of screening tools that neglect to ask about stories of resiliency and positive transformation of the changed SCI body. These ontological spaces are problematic for several reasons; they span the healthcare landscape from primary care to higher education preventing gaps to emerge that can challenge the dominant view, they fail to listen and hence be changed by stories of personal adaptation and transformation of individuals living with stuckness and lastly, future healthcare providers continue to be trained by the medical model embraced by higher learning institutions. By continuing to focus on the stuckness of body parts the epistemological underpinnings of the medical model that reifies the physician as expert is re-affirmed and reproduced. This offers an interesting parallel between the medical model's reluctance to shift mirrored by the SCI patients' inability to move.

That said, small yet important signs of resistance aimed to disrupt such discourses can be noticed, and instead offer a new and positive conceptualization of the immobile body. Anti-narratives seek to reclaim how the whole body is seen and assigned worth as a valued member in society. Despite living in spaces of "in-betweenness", those with a SCI injury are mobilizing within these liminal locations "to make sense of the in-between, to overcome it and leave it behind and with a difference" (Horvath et al. 2015, 40). The differences are taking the form of political advocacy, peer support movements and demands to be a seen as a partner in their care as opposed to a patient. The shared momentum to force healthcare structures to change and respond to their healthcare experiences in a more equitable way supports a sense of community, or "a collective *communitas* where traditional boundaries dissolve" (Downey et al. 2016, 8). Such experiences need to be explicitly heard at all levels, ranging from the practitioner to the educator preparing the next generation of healthcare providers, as a way to move away from a binary conceptualization of the body to one that is more nuanced, recognizing that presence is more than just physicality of form.

References

Banerjea, R., P. A. Findley, B. Smith, T. Findley, and U. Sambamoorthi. 2009. "Co-occurring Medical and Mental Illness and Substance Use Disorders among Veteran Clinic Users with Spinal Cord Injury Patients with Complexities." *Spinal Cord* 47 (11): 789–795. https://doi.org/10.1038/sc.2009.42.

Bonner, Stephen, and Caroline Smith. 2013. "Initial Management of Acute Spinal Cord Injury." *Continuing Education in Anaesthesia Critical Care & Pain* 13 (6): 224–231. https://doi.org/10.1093/bjaceaccp/mkt021.

Cambridge Dictionary Online, s.v. "liminal (*adj.*)." Accessed January 19, 2019. https://dictionary.cambridge.org/dictionary/english/liminal.

Campbell, Catherine, Kerry Scott, Morten Skovdal, Claudius Madanhire, Constance Nyamukapa, and Simon Gregson. 2015. "A Good Patient? How Notions of 'a Good Patient' Affect Patient-Nurse Relationships and ART Adherence in Zimbabwe." *BMC Infectious Diseases* 15 (1): 404–414. https://doi.org/10.1186/s12879-015-1139-x.

Canter, Richard. 2001. "Patients and Medical Power." *British Medical Journal* 323 (7310): 414. https://doi.org/10.1136/bmj.323.7310.414.

Carlson, Licia. 2010. *The Faces of Intellectual Disability*. Bloomington, IN: Indiana University Press.

Cao, Yue, James F. Massaro, James S. Krause, Yuying Chen, and Michael J. Devivo. 2014. "Suicide Mortality After Spinal Cord Injury in the United States: Injury Cohorts Analysis." *Archives of Physical Medicine and Rehabilitation* 95 (2): 230–235. https://doi.org/10.1016/j.apmr.2013.10.007.

Cassell, Eric J. 2011. "Suffering, Whole Person Care, and the Goals of Medicine." In *Whole Person Care: A New Paradigm for the 21st Century*, edited by Tom A. Hutchinson, 9–22. London: Springer International Publishing https://doi.org/10.1007/978-1-4419-9440-0.

Creamer, Deborah Beth. 2009. *Disability and Christian Theology: Embodied Limits and Constructive Possibilities*. Oxford: Oxford University Press.

Disability Rights, and Education Fund. n.d. "Healthcare Stories." Accessed February 22, 2019. https://dredf.org/healthcare-stories/.

Downey, Dara, Ian Kinane, and Elizabeth Parker, eds. 2016. *Landscapes of Liminality: Between Space and Place*. London: Rowman & Littlefield.

Evans, Jeffrey E. 2004. "Why the Medical Model Needs Disability Studies (and Vice-Versa): A Perspective from Rehabilitation Psychology." *Disability Studies Quarterly* 24 (4). https://doi.org/10.18061/dsq.v24i4.893.

Fisher, Pamela, and Dan Goodley. 2007. "The Linear Medical Model of Disability: Mothers of Disabled Babies Resist with Counter-Narratives." *Sociology of Health & Illness* 29 (1): 66–81. https://doi.org/10.1111/j.1467-9566.2007.00518.x.

FitzGerald, Chloë, and Samia Hurst. 2017. "Implicit Bias in Healthcare Professionals: A Systematic Review." *BMC Medical Ethics* 18: 19. https://doi.org/10.1186/s12910-017-0179-8.

Hage, Ghassan. 2009. *Waiting*. Melbourne: Melbourne University Press.

Horvath, Agnes, Bjørn Thomassen, and Harald Wydra, eds. 2015. *Breaking Boundaries: Varieties of Liminality*. New York: Berghahn Books.

Hutchinson, Tom A. eds. 2017. *Whole Person Care: A New Paradigm for the 21st Century*. New York: Springer.

Johnstone, David. 2012. *An Introduction to Disability Studies*. London: Routledge and Taylor & Francis. https://doi.org/10.4324/9780203462379.

Joseph, Meera, Sujen Saravanabavan, and Jeff Nisker. 2018. "Physicians' Perceptions of Barriers to Equal Access to Reproductive Health Promotion for Women with Mobility Impairment." *Canadian Journal of Disability Studies* 7 (1): 62–100. https://doi.org/10.15353/cjds.v7i1.403.

Kasser, Susan L., and Rebecca K. Lytle. 2005. *Inclusive Physical Activity: A Lifetime of Opportunities*. Champaign, IL: Human Kinetics.

Krueger, H., V. K. Noonan, D. Williams, L. M. Trenaman, and C. S. Rivers. 2013. "The Influence of Depression on Physical Complications in Spinal Cord Injury: Behavioral Mechanisms and Health-Care Implications." *Spinal Cord* 51 (4): 260–266. https://doi.org/10.1038/sc.2013.3.

Lim, Sher-Wei, Yow-Ling Shiue, Chung-Han Ho, Shou-Chun Yu, Pei-Hsin Kao, Jhi-Joung Wang, and Jinn-Rung Kuo. 2017. "Anxiety and Depression in Patients with Traumatic Spinal Cord Injury: A Nationwide Population-Based Cohort Study." *PLoS One* 12 (1): e0169623. https://doi.org/10.1371/journal.pone.0169623.

McMillan, Colleen, Joseph Lee, James Milligan, Loretta M. Hillier, and Craig Bauman. 2016. "Physician Perspectives on Care of Individuals with Severe Mobility Impairments in Primary Care in Southwestern Ontario, Canada." *Health and Social Care in the Community* 24 (4): 463–472. https://doi.org/10.1111/hsc.12228.

McMillan, Colleen, Joe Lee, Loretta Hillier, James Milligan, Linda Lee, Craig Bauman, Michelle Ferguson and Kay Weber, 2020. "The Value in Mental Health Screening for Individuals with Spinal Cord Injury: What Patients Tell Us." *Archives of Rehabilitation Research and Clinical Translation* 2 (1): 100032. https://doi.org/10.1016/j.arrct.2019.100032.

Olkin, Rhoda. 1999. *What Psychotherapists Should Know About Disability*. New York: Guilford Press.

Rogers, Brant, Michael Christopher, Zeynep Sunbay-Bilgen, Marie Pielage, Hui-Ning Fung, Lauren Dahl, Jennifer Scott, et al. 2013. "Mindfulness in Participatory Medicine: Context and Relevance." *Journal of Participatory Medicine* 5 (7, February). https://participatorymedicine.org/journal/evidence/reviews/2013/02/14/mindfulness-in-participatory-medicine-context-and-relevance/.

Sezer, Nebahat, Selami Akkuş, and Fatma Gülçin Uğurlu. 2015. "Chronic Complications of Spinal Cord Injury." *World Journal of Orthopedics* 6 (1): 24–33. https://doi.org/10.5312/wjo.v6.i1.24.

Shakespeare, Tom, Lisa I. Iezzoni, and Nora E. Groce. 2009. "Disability and the Training of Health Professionals." *Lancet* 374 (9704): 1815–1816. https://doi.org/10.1016/S0140-6736(09)62050-X.

Smith, Diane L. 2009. "Disparities in Patient-Physician Communication for Persons with a Disability from the 2006 Medical Expenditure Panel Survey (MEPS)." *Disability and Health Journal* 2 (4): 206–215. https://doi.org/10.1016/j.dhjo.2009.06.002.

Stewart, Moira, Judith Belle Brown, Wayne Weston, Ian R. McWhinney, Carol L. McWilliam, and Thomas Freeman. 2013. *Patient-Centered Medicine: Transforming the Clinical Method*. 3rd ed. London: CRC Press and Taylor & Francis.

Sweis, Rochelle, and José Biller. 2017. "Systemic Complications of Spinal Cord Injury." *Current Neurology and Neuroscience Reports* 17: 8. https://doi.org/10.1007/s11910-017-0715-4.

Thomas, David, and Honor Woods. 2003. *Working with People with Learning Disabilities: Theory and Practice.* London: Jessica Kingsley Publishers.
Thomassen, Bjorn. 2015. *Liminality and the Modern: Living Through the In-Between.* London: Routledge.

8

Embodied Perceptions of Immobility After Stroke

Hannah Stott

Introduction

This chapter examines how immobility caused by neurological conditions may be experienced in chronic illness; examining how immobile bodily processes may limit the ability to enact psychological and social opportunities. Immobility in neurological conditions may evolve gradually, slowly eroding the potential activity of the individual, or it may appear suddenly, out-of-the-blue. In conditions such as motor neurone disease, Parkinson's or multiple sclerosis, the point of diagnosis may feel like a bombshell as the individual is confronted with a potential future of immobility and life may feel like it's over (Pretorius and Joubert 2014, 6; Sakellariou et al. 2013, 1770–1771; Smith and Shaw 2017, 16). In conditions such as stroke or brain injury, immobility is presented as a sudden, startling reality. This may feel cataclysmic and sit in stark

H. Stott (✉)
University of the West of England, Bristol, UK
e-mail: hannah3.stott@uwe.ac.uk

contrast with the remembered mobility of the past and a now uncertain future. The abrupt and unexpected nature of immobility can create a unique disruption to how the relationship between the body and self is experienced and conceptualized which has implications for how the individual is able to navigate the social environment (Faircloth et al. 2005, 933–936; Vigh 2009, 425). These themes will be explored later in the chapter, against the backdrop of a recent phenomenological study detailing the (im)mobility experiences of 16 stroke survivors. To pick apart how (im)mobility is experienced in chronic illness, this chapter will provide focus on the individual experience of physiological processes of immobility, linking this to the self, identity, social navigation and existential mobility (Browning and Joenniemi 2017, 40; Janssens 2018, 7; Vigh 2009, 425).

Motility and Exploring the Environment

To deconstruct the experience of losing one's ability to move and use the body, it is useful to begin by pulling apart the lived experience of motility, or in other words the individuals' potential for mobility. Motility is the sense that one is able to act, with purpose, within their environment. Motility draws together one's ability to reflect on life's possibilities and then convert this intangible potential into goal-directed action through the medium of the body. It encompasses the competence of an individual and how able they are to access and interpret their environment and its embedded sociocultural structures (Kaufmann et al. 2004, 750). Understanding this process in terms of the interaction between body and psychosocial processes is useful to shed light on moments motility is disrupted following neurological injury.

Whilst it is true that a key rationale for engaging in meaningful movement is to respond to meet our social needs (Vigh 2009, 425), we also move to meet our physiological needs. As such, our bodies are in constant dialogue with our environments. "*We see chairs that offer up the prospect of rest, food that may be eaten, a cold rain that bids us stay inside. The sensory world thus involves a constant reference to our possibilities of active response*"

(Leder 1990, 18). In a healthy body, these actions remain largely unconscious. When walking, we ignore the corporeal experience of the physical sensations of muscle tension and release, foot placement and sensation of the ground underfoot and we are given cognitive space to focus on the goal of our movement; the meaning it might place on our imagined future (Vigh 2009, 425). Motility is therefore hinged on our ability to silently perceive the physical sensations of our bodies in our environments. Certainty in these perceptual sensations provide a cognitive space to consider how the potential and imagined use of this movement may be meaningful and purposeful, thus allowing movement to underpin who we are and why we act. Though these body processes may seem distant from social models of disability, they are fundamental to developing our understanding of the "*embodied experience of action and meaning making in physical, interpersonal and social settings*" (Papadimitriou 2008, 693).

Phenomenology has contributed much to our understanding of the perceptual experience of motility. Through locating the body at the centre of the lived experience the literature has disentangled the minute experiential detail of our perceptual and sensory processes, and proposed theory as to how and why we respond to our physical environments (Leder 1990, 11–36; Merleau-Ponty 2001, 67–207). This reasoning suggests that whilst we are undoubtedly driven to act by social determinants, which we are able to cognitively reflect on, our environments may also generate movement and action at a more physiological or visceral level (Sakson-Obada and Wycisk 2015, 83).

> In the normal person, every event related to movement or sense of touch causes consciousness to put up a host of intentions which run from the body as the centre of a potential action either towards the body itself or towards the object. (Merleau-Ponty, 2001, 125)

As such, the body becomes situated as the facilitator between the perceptual processes that interpret the environment and as the primary driver for acting out the desires of the individual. We absorb the possibilities of our environment and are sparked into action (Leder 1990, 45–46).

Whilst the body is the "*vehicle for being in the world*" (Merleau-Ponty 2001, 82), its processes for healthy individuals are largely pre-reflective and unremarkable. The body silently gathers the perceptual stimuli which enables us to feel, sense, act and make decisions about the world around us (Azañón et al. 2016, 2; Dieguez and Lopez 2017, 198). For the most part, these processes whir on in our perceptual backdrop, receding out of our conscious awareness. They may only call for our attention when we need to respond and remedy unpleasant bodily sensations. Imagine, striding purposefully across a room and the jarring, perceptual surprise of stepping on a piece of Lego. Suddenly our body demands our attention and we instantly forget our cognitive reasoning for walking in the first place. The advantage of processing our world so silently is we are given the space to reflect on our cognitive drivers for movement—the social actions which define our sense of identity and self (Browning and Joenniemi 2017, 41). The experience of responding to perceptual sensations is learnt, temporal and constantly evolving. We learn from previous moments in which the body was navigated successfully and adapt our response accordingly. Navigating the body is thereby shaped by past experiences, present needs and imagined future identities, making social mobility physiologically embodied (Cresswell 2006, 57–59; Veal 2017, 307).

Neurological Injury and Disrupted Motility

Neurological injury and chronic illness provide moments to examine disrupted motility. This type of disruption may be experienced in moments the body no longer responds as expected. It is this failure or disruption of perceptual processes which catapults our bodies into our conscious awareness. Suddenly the body is noticeable and can no longer be taken for granted. This changes how we relate to our bodies and the ways directed action is able to form identity (Browning and Joenniemi 2017, 40; Gadow 1980, 182; Sakson-Obada and Wycisk 2015, 83). If this disruption affects how the body moves, the deficit becomes observable to a public gaze which has implications for performative and social identities. It becomes suddenly necessary to bring conscious focus to the body to attempt to remedy the inaction. This body obstacle, inhibiting the path of action and utility, is mirrored in the invisible microcosm of perceptual processing. Inside the body, the transmission of environmental information through the highway of our central and peripheral nervous system become lost in transit, stuck, unable to travel their usual pathways and be translated into action.

Apraxia is a condition in which damage to the brain results in an inability to act out purposeful actions disrupting how the body, objects and the environment are navigated. Apraxic patients have described being able to locate and hold the tools necessary to achieve daily activities but as being unable to sequence the body actions to execute the manoeuvre (Arntzen and Elstad 2013, 64). Frustratingly, patients may have insight into their perceived failure to act but be unable to correct the body response. For this person, it takes the visual cue of seeing the failed movement to realize cognitively what has gone wrong, thereby making the relationship with the body and action conscious rather than perceptual.

The relationship between the body, its interaction with the environment and our sense of identity is built upon a sense of ontological trust that our body senses will provide us with accurate information to interpret our surroundings (Giddens 1991, 38). Having embodied experiences of trust, security and self-confidence are fundamental to feeling empowered to participate in society (Gyllensten et al. 2010, 442). Yet, moments in which the body fails to act may produce conflicting emotional responses: perhaps confusion, embarrassment or anxiety. Having to focus and remedy the body uses up the cognitive space which may previously have focused on individual intentions and needs. The physicality of the perceptual absence is therefore able to permeate all areas of identity. These emotional responses create limitations on what can be imagined, which play out in terms of social mobility (Fig. 8.1). Examining how the bodily sense of (im)mobility affects psychosocial components of identity is key to better understanding the effect of multiple immobilities experienced in illness (Veal 2017, 307).

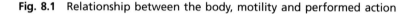

Fig. 8.1 Relationship between the body, motility and performed action

Affective and Social (Im)mobility

The physical perceptual experiences which create a sense that our body is ours allow the body to be purposefully utilized to meet our needs (Imaizumi et al. 2016, 85). Through being able to act out intentions in our environment we are able to demonstrate our purpose, which ties the sense of our bodies to our personal, occupational and relational contexts. This idea of *acting* identity inexorably links movement and action with cognitive and psychological processes. These identities are not fixed but are temporal or flexible. Some identity stories are a thread through an individuals' life and are important to create a sense of a remembered past and an imagined future (Asai et al. 2016, 8). Other threads come and go and can be disrupted by illness and immobility (Ellis-Hill et al. 2008, 154). Being able to perform social actions is a contributor to the plural identities which serve to maintain the biographical narrative of the self (Browning and Joenniemi 2017, 41; Giddens 1991, 53). It is through the loss of autonomous action that it becomes possible to trace how our body sense relates to the sense of who we are. After all, how can we be valuable to ourselves, our family and society if we cannot move, act and contribute?

Imagination facilitates the conceptualization of opportunities and how experiences are made sense of, which are developed in relation to social meanings and representations (Salazar 2012, 864). Yet little is known about how imagination of potential mobility and social activity is affected following neurological injury. A qualitative exploration of imagined movement tasks after stroke suggested that, prior to neurological disruption, individuals focused on imagining more complex activity scenarios, whereas post-stroke, imagined activities were focused on motor outputs such as moving the fingers (Schuster et al. 2012, 12). Participants in this study highlighted how cognitive distractions impinged on their ability to imagine movement which they now found required more concentration, space and time. The authors of this study suggest that imagining activity may provide an opportunity to imagine a 'whole self', which may be important to re-establish the neurological connections necessary for movement (Schuster et al. 2012, 12). The role that imagination plays in the formation of movement, action and the self are not well understood but may be an important area for further research if personal and cultural imaginaries interweave to create meaning for the individual.

It is likely that individual imaginaries are influenced by social contexts. However, disability discourse exists in an ableist context, meaning the immobile body is usually considered as problematic and in need of fixing what it can no longer do (Loja et al. 2013, 191). Sociocultural narratives suggest that disabled or immobile people are incapable or not fully human (Goggin 2016, 535; van der Meide et al. 2018, 2244). This serves to throw off course culturally valued norms of personal identity (Vigh, 2006, 56). For the individual experiencing immobility, the ableist narrative may sit alongside remembered, embodied experiences of a moving body, creating a lived experience which may be difficult to reconcile. Disability identities may therefore be resisted, avoided or ignored. Imagining a future self as whole or healthy may be an adaptive coping strategy to avoid confronting challenging sociocultural norms which prioritize able-bodied futures (Rice et al. 2017, 216). Yet this strategy may be problematic as it is incongruent with accepting and reintegrating the sense of an altered physical body.

However, the reality of the immobile body is impossible to avoid. Moments of bodily *in*action or failure occur during interactions with the self, others and the environment, forcing the individual to recognize their limitation. Sudden exposure to "*landscapes of exclusion*" or spaces associated with disability (e.g. hospitals, rehabilitation groups, disabled toilets or parking spaces) may also contribute to a sense of marginalization (Kitchin and Law 2001, 288). Physical markers of immobility such as walking aids make disability visible to the external gaze and may be mirrored back to the individual in terms of stigma and stereotypes (Papadimitriou 2008, 699). These challenging moments in which disability is experienced publicly create emotional responses in the individual, such as shame. Shame may then silence the communication of these body experiences, thereby further exacerbating isolation and social immobility (Mitchell 2018, 493). Mobility can therefore be construed as situated in emotional, social and relational contexts (Janssens 2018, 1).

Fundamentally, the experience of physical disability changes the course of an imagined life trajectory. It questions the very nature of an individual's existence and identity. In sudden-onset illness, it may cause biographical disruption, in which having to attend to body processes affects the way the body can be understood and the ideas one can hold about oneself (Bury, 1982, 169; Engman 2019, 120). It may feel like "*there is a sense of no return… normal life has come to an end*" (Ellis-Hill et al. 2000, 728). To cope with the effects of living with this kind of cognitive dissonance it is common for those experiencing chronic illness to create a psychological division between their disabled body to protect the sense of who they are or their sense of self (Charmaz 1995, 657; Gadow 1980, 176; Leder 1990, 68). So whilst in health, the self and the body may be considered as inseparable, in illness, the lived body becomes noticeable and a site of constraint. From this tension, the body is experienced as an obstacle and becomes objectified. "*The 'object' body emerges out of the primary unity of the lived body in the experience of feeling encumbered by oneself*" (Gadow 1980, 176). This idea of separation between the body and self has been described in people with motor neurone disease, amongst many other conditions (Sakellariou et al. 2013, 1771). In this condition, individuals experience the body as uncontrollable, unreliable and untrustworthy. This physiological experience makes the body consciously noticeable and creates the need to create psychological distance between the body and self. The uncertain disease trajectory also affects imagined future physical and social mobility, so individuals may attempt to exert control over their lives by developing a preferred idea of self which may not be so reliant on previous markers of identity (Sakellariou et al. 2013, 1771).

Body Perception and (Im)mobility After Stroke

Stroke creates a sudden and shocking change to how the body is experienced, felt, perceived and manoeuvred. It is the leading cause of severe and complex disability in the UK (Adamson et al. 2004, 174). One study of 63 patients in which the majority had experienced mild stroke showed that 80% experienced problems with mobility (Kristensen et al. 2016, 1). The body changes experienced after stroke are diverse and affect mobility in differing ways. Muscle weakness affecting one side of the body is a contributor to reduced function and affects around three quarters of stroke survivors (Canning et al. 2004, 304; Lawrence et al. 2001, 1281). Changes to the ability to interpret sensations, awareness of the body and the environment, vision and balance also affect the ability to move (Lawrence et al. 2001, 1281; Nijboer et al. 2013, 3; Tyson et al. 2008, 168).

These physical changes to body perception may be perceived by stroke survivors as being at the core of their experience: "*The arm and leg weakness was the stroke*" (Doolittle, 1991, 236). These perceptual experiences may alienate the individual from the sense of their body. Stroke survivors have described experiencing limbs feeling "*as if a log of wood were hanging from [the shoulder]*" (Head and Holmes, 1911, 224); "*dead*", "*gone*", or "*cut off*" (Doyle et al. 2014, 996). They may experience phantom sensations such as feeling that a limb is not in its actual place or is moving in a way that is not real (Antoniello et al. 2010, 1117). The bizarreness of such bodily perceptions has led stroke survivors to feel like their bodies are strange or unfamiliar and they have described feelings of alienation (Ellis-Hill et al. 2000, 728; Faircloth et al. 2005, 933–936; Stott 2019, 38–40; Tham et al. 2000, 401). These changes to physical body perception processes have ramifications for how the body is able to function and navigate the environment and implications for the psychosocial lived experience (Rubio and Van Deusen 1995, 552).

(Im)mobility after stroke is usually conceptualized by health professionals in terms of activity, and patients' progress is determined via measuring how well they move or utilize the body (Eng et al. 2002, 756). The premise for this is that mobility is an important precursor for independent living which may in turn reduce the economic burden of care after stroke. As such, common rehabilitation goals purport that stroke survivors should "*Use it or lose it*" (Allred et al. 2014, 1). This kind of rehabilitation discourse puts the onus on stroke survivors to adopt a mind over matter approach to achieve bodily mobility through using repetitive practice of movements to develop their neural pathways. This discourse is intended to provide motivation and hope to the stroke survivor living with the experience of immobility. However, it may not fully consider what it feels like to attempt to engage with a body that feels unfamiliar and provides incorrect information which is hard to make sense of. It is therefore useful to develop our understanding about how disrupted motility and mobility is experienced by the individual at both physiological and psychosocial levels, and how these conceptions may interact during attempts to reanimate the immobile body.

Experiences of Immobility After Stroke

To explore the ways in which sudden and chronic immobility alter the perceptual experience of the lived body, this chapter draws on qualitative phenomenological research exploring how the immobile body was experienced by a group of stroke survivors. The physical and psychosocial body perception experiences were collated from a diverse group of sixteen stroke survivors, living in the South West of England. Individuals were able to communicate verbally and were interviewed in their own homes.

Most participants were mobile, albeit in a more limited sense than before their strokes and were able to move using sticks or walking frames. One person used a wheelchair and another an electric scooter when she was outside.

Physical Perceptions of Immobility After Stroke

'I'm Stuck'

Becky was 58 years old, and had had 3 strokes, the most recent 9 years prior to our conversation. She lived alone in a small adapted bungalow and used a walker at times to mobilize indoors and an electric scooter outside. She described feeling as if her right leg was "*amputated*" from the knee down and her right side felt "*like it doesn't belong to you… it's weird but it don't exist*". She was resilient, optimistic and motivated to access and engage with the world around her. She described her sense of immobility and fear of being physically stuck as she attempted to alight from a coach when on a recent excursion.

> Everybody else just… goes out grabs the head [rest] thing, gets up and walks out… whereas my first reaction is, 'Help I can't move'… 'I'm stuck' … but I'm not, no more than they are… it just feels that way… [the body] doesn't just get yourself up and get yourself out… it takes longer to do those movements. (Becky, 58, 9 years post-stroke)

For Becky, the pause between her intention to move and the actualization of that movement made her feel panicked and helpless. Stuart described this perception in more detail.

> I feel like I ought to be able to do everything I could do…and just suddenly [the feeling] it's not there on the right side… so it's that gap, that's the thing… [the feeling of the body not being there means] you can't find…anywhere in your brain, the signal that will move the arm… it feels like it should be able to do it, but somehow it just can't… in terms of the brain, you know if you knock out a portion of it… which

unconsciously moves my arm or my leg and everything else, it's just not there. (Stuart, 60, 20 months post-stroke)

Stuart was 60 years old and had been forced to retire 20 months previously due to his stroke. He felt like his body was missing on his right-hand side—a feeling he closely associated with the feeling of not being able to identify how to make the body respond. He walked outside with a stick. He found that when his body was static it felt 'normal' and comfortable. However, when he attempted to move, the bodily absence became present or conscious and effortful. The conflict between body and mind became apparent: the static body was filled with a sense of action or purpose, yet when the body was urged to move, its absence was made obvious—the mind perceived the body as effortful and a poor mimic of a 'normal' body. For Stuart, this gap in response was a source of discomfort.

Experiences of feeling stuck due to a lack of bodily response are rarely described in the literature, and little is understood about how this perception is managed or how it affects the individual as a whole. This series of interviews suggested that it was the surprise that the unresponsive body would not meet the needs of the individual which was discomforting; the disconnect between cognitive intention and the ability to enact an action. "*I'm not in discomfort if you see what I mean… I'm only in discomfort when I can't move it…*" (Marc, 64, 2 years post-stroke).

Stroke research has shown that it takes time to realize the effects of body changes after stroke, and in the early days particularly, stroke survivors may forget or not realize their body limitations (Blijlevens et al. 2009, 470; Faircloth et al. 2004, 78; Tham et al. 2000, 402). It is often through attempting and failing in activities that individuals learn to recognize their new-found limitations. Mistakes are therefore a key component to rediscovering the body. In a separate study, this was illustrated in a moment in which a woman struggled to perceive her arm, but did not recall her body limitations: "*she unconsciously used her paralyzed arm as before the stroke, with the result that the flower vase was thrown to the floor*" (Kvigne and Kirkevold 2003, 402). These failed actions may occur because changes to body perception which commonly occur after stroke such as numbness, muscle weakness, paralysis or being unable to direct attention to a body area are sensations which are not physically perceived when the body is not moving. They are therefore easy to forget. The effect of this, particularly in the early stages post-stroke, is that moments of bodily immobility may be surprising as the impact on how the body is used becomes noticed and learned.

'A Dead Weight'

Participants highlighted how changes to the sense of the body size, weight or shape affected their ability to mobilize. In these instances, the body was not altogether unresponsive, but the response was cumbersome, a burden and hard to manage. Heaviness of the limbs felt like "*a lump of lead*" (Sarah); like having a "*young child grabbing hold of your knee*" (Ali); like carrying "*your baby*" (Leah); like "*dragging a 10 lb dumbbell behind you*" (Tom); like having the leg "*literally stuck to the floor*" (Johan); like "*wearing divers' boots*" (Becky); and like "*a dead weight*" (Michelle). Moving the heavy limb required so much "*perseverance… to make it go…It probably does use up more calories*" (Sarah).

> I will have been walking as well as I can all day and then suddenly I'm really crippled again… suddenly I can't bend my leg and suddenly it weighs a ton…the muscles in the knee seizing up and then as the muscles in the hip and the thigh get tired then the leaden feeling comes on… strengthening the muscles and improving my stamina gets round the leaden feeling, it doesn't stop the stiffness necessarily… but I understand now that the heaviness is what I feel because the muscles are weak or tired… and they feel like they're pulling up vast weights. (Johan, 66, 11 months post-stroke)

The physical sensation of heavy limbs was, for many, an accepted consequence of stroke, yet the impact of this sensation on mobility was often surprising and limited activities. Like the experience of being 'stuck', the sensation was made conscious when the body could not be taken for granted and when it impeded the desired action.

> I do collapse, like the undercarriage has given way…When I'm walking and it happens, I am saying 'I will walk on', but my legs are saying 'if you want to walk on, carry on without us' [laughs]… That's exactly how it feels… 'You might want to but we don't'. (Daniel, 79, 11 months post-stroke)

Daniel was tremendously isolated by his unreliable body and had experienced collapses several times when he was out alone. This had now made him fearful of leaving the house. He described his experience through the dichotomy of the body and mind, envisaging them as two opposing forces. For Daniel, his immobility was experienced as problematic through a range of interrelated experiences which created a cumulative impact. His experience of physical collapse led to fear of mobilizing his untrustworthy body, which caused isolation. He found his experience difficult to make sense of and felt rejected by health professionals, meaning he struggled to imagine a future where he was mobile. Literally and figuratively he had nowhere left to go.

Like Daniel, many participants engaged in dialogue to engage the object body in movement.

> I'm having a conversation with the person I call Jiminy Cricket…he's on my left shoulder…he's a little green cricket…he's there all the time…it's him that's saying 'move your left leg' and stuff…and then 'right you're going now, just keep going' you know 'eyes front and keep going'…he's an imaginary person obviously…he never answers back… you don't have a conversation with him, he just says 'do it' and you do it [laughs]… well that feels worse than the shoulder feels… because you've got no control over him… he's just a figment of your brain… sort of gets on your nerves. (Marc, 64, 2 years post-stroke)

It is likely that changes in body sensations which may make a limb feel alien and disconnected from the whole lead to limbs becoming objectified by the individual. Stroke survivors describe this as a lost *"sense of bodily wholeness"* (Doolittle 1991, 237) or as feeling *"trapped in an object like body"* (Ellis-Hill et al. 2000, 729). This physical feeling becomes reflected in language and the body is frequently referred to in the third person (Tham et al. 2000, 401). It is the experience of mobilizing the object body which can be thought to create this dialogue between the mind and body (Faircloth et al. 2005, 933–936). This experience of altered physical perceptual processes is lived through experiences more closely tied to identity—that the self may be able to exert control over an uncontrollable object body. Interestingly, as in Daniel's account above, the nature of this dialogue may also be at times bidirectional, as the body talks back and refuses to cooperate with the will of the mind (Kirkevold et al. 2012, 667). This interaction shines light on the complex ways that lost perception and actions can influence post-stroke ideas about identities and self which may be key factors for personal and social mobility.

Psychological Immobility After Stroke

The idea of the self being at odds with the body creates two actors in a scene in which previously there had been only one. The body becomes a hindrance, which is now unable to perform the intentions of the self (Gadow 1980, 178). Psychologically the individual must reconcile the loss of their previous life and being thrust into a new disability identity, which may be perceived by the individual and society as a stigmatized, undesirable social out-group (Bogart 2014, 108). Experiencing the body as separate to the self can be helpful because it allows the frustration of immobility to be directed to *another*. Though this may protect the self to an extent, over time, the effect of lost personal and social mobility will eventually be noticed and will have to be reconciled.

'It's a Battle'

Living with the unresponsive body was repeatedly described across all accounts, as "*sheer frustration*", "*a waste of time*", "*useless*", "*difficult*", "*isolated*", "*terrible*" and "*weird*". For many, this led to feelings of anger and low mood. The body's refusal to cooperate caused a complex sense of disconnection between the body and mind.

> Your mind is being taken up with some other things and because it's being taken up with other things the leg doesn't want to bend... so it then becomes a bit of a battle and it then becomes very restrictive... it does become a pain in the neck... that can just be very frustrating... the mental mindset then gets irritated by it... which then just makes the problem worse... because the more irritated you get the more it won't work. (Toby, 46, 19 months post-stroke)

Toby, a young and very active professional, was used to being in control of his life. Post-stroke he experienced limited movement on his right side which set the relationship between his body and mind in conflict. To cope he engaged in hours of self-directed rehabilitation which provided a sense of independent control over his battle with the rebellious, object body.

Joel, a self-confessed 'negative thinker', articulated with clarity how an absence of physical response affected his mood. His loss of automatic action forced unwanted reflection on the body and consequently the meaning of what had been lost. Momentary gaps in response highlighted existential fears about the future. His emotional reaction showed how his loss of body function and utility was tied to ideas about who he was and who he might become.

> You need to... just pull your laces and it just doesn't happen. ... you think you would automatically do it without thinking. ...But you have to think and it doesn't work... [it feels] [pause] frustrating, frightening and puzzling... [it makes me think] 'Is this ever going to get any better?'... 'Why?'...So it's just frustration really... You get angry... 'I can't do it'. (Joel, 64, 2 years post-stroke)

For Joel, the meaning of his physical immobility meant he experienced fear, frustration and uncertainty about the future. These psychological interpretations of the physically immobile body created existential questions about identity and purpose, which in turn reduced his confidence and social mobility. These physical and psychological changes created a sense of *stuckedness* (Pettit and Ruijtenberg 2019, 1). A similar sense of existential (im)mobility has been noted in individuals experiencing limb amputation who described the loss of space in which to perform actions and freedom as dominant contributors to this psychological experience (Norlyk et al. 2013, 7). Being able to distinguish between the loss of bodily function and one's sense of self-worth has been shown to be important factors when accepting changes to function or movement, and those who perceive themselves as useless are more likely to experience symptoms of depression (Townend et al. 2010, 561).

Coping with the uselessness of the body led a few participants to consider amputation to get rid of the offending body areas.

> It's all a waste of time this arm now… because it won't work, it's there but it won't work [laughs]… I often think 'well should I just cut it off one day'… get it out the way… but then I think, well my shoulders still there so… [it makes me feel] bad. (Marc, 64, 2 years post-stroke)

Marc experienced limited movement and awareness of his body down his left side and was no longer able to work. He lived with his wife and was able to leave the house independently, walking with a stick. His remarks highlighted how his view of his body was primarily about functionality. His description of amputating the limb demonstrated how intense the experience of bodily objectification can become after stroke—a response to the loss of physical sensation and sense of body wholeness, combined with a psychological response to the loss of utility and the impact of this on identity.

Social Immobility After Stroke

After stroke, immobility is often visible and the body may feel conspicuous (Kvigne and Kirkevold 2003, 1303). Mobility may be facilitated using sticks, walking frames or wheelchairs. Limbs may be visibly uncooperative, held in unnatural angles and movements may be slow or jerky. The physical environment becomes an endless source of challenges and the moving body may struggle to recognize and unconsciously respond to uneven floors, fluffy carpet, imperceptible steps or obstacles in a path. Body changes which may be troublesome at a personal level become perceived in relation to the consequences of having to engage with the physical environment, thereby making fear of falling a main constraint on mobility. This fear was iterated by many participants.

> I no longer have a correct sense of what's vertical... I need to be slightly off-balance in order to be vertical... and I'm therefore afraid I'm going to fall over... that leads to fear and anxiety... so that slows me down... walking across a carpark or something is terrifying... because... I'm in free space... (Johan, 66, 11 months post-stroke)

Living with the (im)mobile body in an external environment exposes individuals more overtly to sociocultural ableist narratives which push individuals into stereotyped out-groups of disability (Daruwalla and Darcy 2005, 553; Loja et al. 2013, 193). For participants in this study, being the recipient of those stereotyped attitudes at an individual level led some to reduce their mobility and social participation in response.

> [if people see me with my hands clenched] they might think 'there's something wrong with you'... 'she's not thinking... she doesn't look normal... could be anything wrong with her'... sometimes it makes you feel that you think they're talking about you... [you want] them to realise that... everybody's not normal... sometimes you feel a bit sad 'cause they don't realise that you've had a stroke... it affects me going out... I'd never go in town on my own now... because people might look at you all the time... (Michelle, 49, 4 years post-stroke)

Participants described a feeling of being ostracized by society due to their disability. They became perceived by others primarily as their body impairment as opposed to who they were. "*I'm fascinated that people will... treat you as an object*". (Johan, 66, 11 months post-stroke). Becky elucidated this comment further by describing how the loss of social mobility after stroke dehumanized the individual and questioned the very essence of identity in the eyes of the observer.

> [In society] how people see you is important... it's like you fall off the ladder... you just go zoomp, to the bottom... society as a whole, it doesn't value anybody who is different, do they? So, if you're different you don't fit in... you kind of slide down the scale... Sometimes [I feel]

angry… like being punished for something you didn't do… and you can't fight back… you don't exist, you're not a person, you're just almost like I dunno, a digit… I'm human… I'm in here, I want to shout that sometimes… [other people] they can't see past what they can see. (Becky, 58, 9 years post-stroke)

If mobility is intrinsically linked to the opportunity to accrue financial, social and cultural capital, then physical disability removes the autonomy to accrue these contributors to status and identity (Salazar and Smart 2011, ii). Being unable to climb the ladder provides a conceptual block in the road after stroke, which may affect the sense of identity and self. The experience of feeling alienated from society and suddenly becoming a part of a stigmatized out-group is hard to reconcile. It may reinforce the sense the body is 'not me' and create further distance between the body and the self (Kitzmüller et al. 2013, 24). Yet how this interacts or validates an already shattered physiological sense of bodily wholeness is not known. It is possible that interpretations of social representations of ableism may intermingle with physiological and psychological perceptions of physical loss to cumulatively impact social (im)mobility.

Immobility and Communication After Stroke

Communication is embodied. Bodies are our anchor to the world and filter our perceptual and visceral experience. It is from this corporeal experience that thoughts are generated, meaning is made, narratives formed and experiences verbalized. It is "*bodies that speak, where speech represents the struggle to make sense of experience and to communicate that experience to oneself self-reflexively as well as to others*" (Murray and Holmes 2014, 15). So how does the experience of bodily strangeness after stroke affect one's ability to trust, make sense of and communicate ideas about the body and self?

In this study, participants frequently found the body hard to make sense of, meaning it was difficult to trust and rely on their corporeal experience. Alistair reflected on his sensations of stiffness and pain in his leg.

> Nobody has explained to me, if you like, the physiology of 'why does it feels as if it needs oiling? What is it?'... So, I look for signs of improvement... I'm hoping if I just keep working at it, I don't know, maybe I shall get the movement back... (Alistair, 72, 11 months post-stroke)

In the absence of understanding his body perceptions, Alistair, like many others, tried his best to keep moving and retained hope his body perception would improve. However, without understanding he could not choose how best to direct his actions to meet his goals, and he was left with uncertainty about his potential future mobility.

To make sense of and remedy the body after stroke, one must be able to conceptualize and communicate those changes verbally. However, novel and unusual body sensations may sit beyond the realms of common language and may only be described in relation to failed movements (Connell et al. 2014, 153). Difficulty communicating may be compounded as 21 to 38% of stroke survivors experience changes to how they are able to produce language (Berthier 2005, 164). All participants in this study showed difficulty finding the words to convey the physical and psychological experiences of their altered bodies. Some used metaphors, and described feeling like the '*Creature from the black lagoon*', '*Edward Scissorhands*' or like they were '*trapped in a bag of sand*'. They seemed embarrassed using these terms and laughed at themselves or dismissed their sensations as just '*weird*'.

Communicating the nature of bizarre sensations, such as feeling that limbs are missing or moving on their own, may not feel socially acceptable and may be avoided in conversation for fear of being perceived as mentally unstable (Antoniello et al. 2010, 1119; Klinke et al. 2015, 1633). This issue was highlighted by Marc, who kept his thoughts about limb amputation private from professionals.

> No [I wouldn't talk to doctors]... I could go on and on and on for days and days, I could spend a whole day with them... just sitting there whinging and moaning about my body... and that is something definitely I can't do... well I'll be diagnosed as being schizophrenic or something... or depressed... they'd think I'd had a screw loose or something... I don't want to be labelled with that. (Marc, 64, 2 years post-stroke)

A lack of confidence in communicating after stroke has been shown to limit engagement with health professionals (Eames et al. 2010, 74). This has implications for personal and social mobility as stroke survivors may be limited in what health advice and intervention they are able to access. This means that problematic ideas about the body such as feelings that limbs are disconnected or should be removed may not be discussed and normalized. Limited access to knowledge may block understanding of body processes which could affect rehabilitation goals and contribute to psychological experiences of *stuckedness*. Sharing knowledge may be key to developing ideas about the self and identity, which in turn ensures flow of ideas in relationships, the building of networks and access to social capital (Kaufmann et al. 2004, 750). As such static or immobile ideas about the body have implications for how an individual is able move along their recovery pathway in complex biopsychosocial ways.

Conclusion

In this chapter, stroke survivors' accounts showed the range of embodied domains in which mobility may be impacted and how these can interact within individual experience (Fig. 8.2). Participants experienced changes to a range of physiological perceptions which affected how they physically perceived the body and in turn how the body responded to their needs. Many felt disembodied and attempted to assert control over the object body using conscious dialogue. The static body felt unremarkable, yet awareness was brought to altered perceptions such as limb heaviness, absence, pain or unresponsiveness when participants attempted to mobilize. The immobile body was physically uncomfortable, socially inhibiting and caused frustration and conflict. This created a sense of psychological separation between the body and self and impacted identity. Difficulties engaging in social experiences validated these psychological constructs. Participants described isolation due to their uncontrollable, unresponsive and untrustworthy bodies. This was further problematized when participants attempted to verbally convey their experience but could not easily access the words or the commonality of a frame of reference to explain their unusual body experiences.

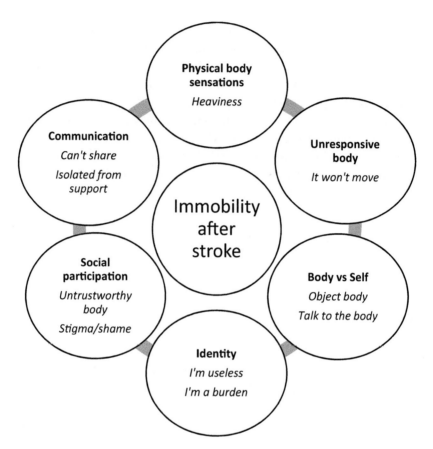

Fig. 8.2 Multifaceted experiences of immobility after stroke

These embodied experiences of immobility showed how the experience of altered physiological sensation can constrain how the individual is able to build ideas of identity and self which are key facets of social mobility. The experience of disrupted mobility is a key theoretical construct linking neurological deficit to the more tangible and visually observable effects on behavioural and social mobility. It is through understanding the experience of disrupted mobility after stroke that it is possible to explore the perceptual, psychological, social and environmental domains which affect mobility more generally.

The experience of losing the ability to perceive and interpret physiological processes after stroke shines a light on the perceptual information which is usually processed silently and unconsciously. It is through taking a closer look at these mechanistic drivers between mind and body that we can understand more about the human experience and adjustment in health and illness. One growing area of research in this area is examining the benefits of embodied therapeutic techniques such as mindfulness. This technique encourages conscious bodily attention as a tool to strengthen metacognitive awareness and has demonstrated efficacy as a treatment for many health conditions. For example, it has been shown to improve physical function, psychological well-being and acceptance in chronic pain (la Cour and Petersen 2015, 649) and to reduce blood pressure and psychological distress in patients with cardiovascular disease (Scott-Sheldon et al. 2019, 5). Mindfulness techniques are also being explored as ways to optimize sport performance in healthy individuals, through enhancing cognitive control and attention regulation (Crivelli et al. 2019, 85). It is possible that these techniques could be applied within general populations who also experience a constantly evolving and changing body experience through ageing, injury, illness, changes in body weight, fatigue, pregnancy or menopause. At present, many of these perceptual changes are internalized, unobserved and silent, yet through improving recognition of body processes, it may be possible to affect positive experiences and improve mobility. Further research is required to move forward from a dualistic notion of the mind and body, to a truly multimodal and holistic notion of body perception.

Learning from experiences of physiological perceptual loss such as stroke emphasizes the key facets of how changes to body processes, action and mobility impact upon our interaction within personal and social environments. It is then possible to see how a better understanding of our multimodal body experience could influence how we design public and domestic spaces to meet our bodily and psychosocial needs more sensitively during the times our bodies become noticeable or less mobile in social environments. Developing public environments which are supportive rather than prohibitive to our body experience could improve how comfortable we feel manoeuvring ourselves in these domains which interacts with how we are able to define our identities through these actions. As such, the improvement of urban spaces is recognized as an important issue considering our ageing population and the increased need to create active and inclusive spaces (Cinderby et al. 2018, 410). Visibly inclusive spaces would also challenge ableist norms and therefore internalized experiences of stigma for those experiencing changes to their mobility. It is through developing a more in-depth understanding of our physiological, psychological and social engagement with (im)mobility that it becomes possible to imagine a more inclusive and holistic understanding of the human experience.

References

Adamson, Joy, Andy Beswick, and Shah Ebrahim. 2004. "Is Stroke the Most Common Cause of Disability?" *Journal of Stroke and Cerebrovascular Diseases* 13 (4): 171–177.

Allred, Rachel P., Soo Young Kim, and Theresa A. Jones. 2014. "Use It and/or Lose It—Experience Effects on Brain Remodeling Across Time After Stroke." *Frontiers in Human Neuroscience*. https://doi.org/10.3389/fnhum.2014.00379.

Antoniello, Daniel, Benzi M. Kluger, Daniel H. Sahlein, and Kenneth M. Heilman. 2010. "Phantom Limb After Stroke: An Underreported Phenomenon." *Cortex* 46 (9): 1114–1122.

Arntzen, Cathrine, and Ingunn Elstad. 2013. "The Bodily Experience of Apraxia in Everyday Activities: A Phenomenological Study." *Disability and Rehabilitation* 35 (1): 63–72.

Asai, Tomohisa, Noriaki Kanayama, Shu Imaizumi, Shinichi Koyama, and Seiji Kaganoi. 2016. "Development of Embodied Sense of Self Scale (ESSS): Exploring Everyday Experiences Induced by Anomalous Self-representation." *Frontiers in Psychology* 7: 1005. https://doi.org/10.3389/fpsyg.2016.01005.

Azañón, Elena, Luigi Tamè, Angelo Maravita, Sally A. Linkenauger, Elisa R. Ferrè, Ana Tajadura-Jiménez, and Matthew R. Longo. 2016. "Multimodal Contributions to Body Representation." *Multisensory Research* 29 (6–7): 635–661.

Berthier, Marcelo L. 2005. "Poststroke Aphasia: Epidemiology, Pathophysiology and Treatment." *Drugs & Aging* 22 (2): 163–182.

Blijlevens, Heleen, Clare Hocking, and Ann Paddy. 2009. "Rehabilitation of Adults with Dyspraxia: Health Professionals Learning from Patients." *Disability and Rehabilitation* 31 (6): 466–475.

Bogart, Kathleen R. 2014. "The Role of Disability Self-concept in Adaptation to Congenital or Acquired Disability." *Rehabilitation Psychology* 59 (1): 107–115.

Browning, Christopher S., and Pertti Joenniemi. 2017. "Ontological Security, Self-articulation and the Securitization of Identity." *Cooperation and Conflict* 52 (1): 31–47.

Bury, Michael. 1982. "Chronic Illness as Biographical Disruption." *Sociology of Health & Illness* 4 (2): 167–182.

Canning, Colleen G., Louise Ada, Roger Adams, and Nicholas J. O'dwyer. 2004. "Loss of Strength Contributes More to Physical Disability After Stroke Than Loss of Dexterity." *Clinical Rehabilitation* 18 (3): 300–308.

Charmaz, Kathy. 1995. "The Body, Identity, and Self: Adapting to Impairment." *Sociological Quarterly* 36 (4): 657–680.

Cinderby, Steve, Howard Cambridge, Katia Attuyer, Mark Bevan, Karen Croucher, Rose Gilroy, and David Swallow. 2018. "Co-designing Urban Living Solutions to Improve Older People's Mobility and Well-Being." *Journal of Urban Health* 95 (3): 409–422.

Connell, Louise A., Naoimh E. McMahon, and Nicola Adams. 2014. "Stroke Survivors' Experiences of Somatosensory Impairment After Stroke: An Interpretative Phenomenological Analysis." *Physiotherapy* 100 (2): 150–155.

Cresswell, Tim. 2006. "'You Cannot Shake That Shimmie Here': Producing Mobility on the Dance Floor." *Cultural Geographies* 13 (1): 55–77.

Crivelli, Davide, Giulia Fronda, and Michela Balconi. 2019. "Neurocognitive Enhancement Effects of Combined Mindfulness-Neurofeedback Training in Sport." *Neuroscience* 412: 83–93.

Daruwalla, Pheroza, and Simon Darcy. 2005. "Personal and Societal Attitudes to Disability." *Annals of Tourism Research* 32 (3): 549–570.

Dieguez, Sebastian, and Christophe Lopez. 2017. "The Bodily Self: Insights from Clinical and Experimental Research." *Annals of Physical and Rehabilitation Medicine* 60 (3): 198–207.

Doolittle, Nancy D. 1991. "Clinical Ethnography of Lacunar Stroke: Implications for Acute Care." *The Journal of Neuroscience Nursing: Journal of the American Association of Neuroscience Nurses* 23 (4): 235–240.

Doyle, Susan D., Sally Bennett, and Brian Dudgeon. 2014. "Upper Limb Post-stroke Sensory Impairments: The Survivor's Experience." *Disability and Rehabilitation* 36 (12): 993–1000.

Eames, Sally, Tammy Hoffmann, Linda Worrall, and Stephen Read. 2010. "Stroke Patients' and Carers' Perception of Barriers to Accessing Stroke Information." *Topics in Stroke Rehabilitation* 17 (2): 69–78.

Ellis-Hill, Caroline, Sheila Payne, and Christopher Ward. 2000. "Self-body Split: Issues of Identity in Physical Recovery Following a Stroke." *Disability and Rehabilitation* 22 (16): 725–733.

Ellis-Hill, Caroline, Sheila Payne, and Christopher Ward. 2008. "Using Stroke to Explore the Life Thread Model: An Alternative Approach to Understanding Rehabilitation Following an Acquired Disability." *Disability and Rehabilitation* 30 (2): 150–159.

Eng, Janice J., Kelly S. Chu, Andrew S. Dawson, C. Maria Kim, and Katherine E. Hepburn. 2002. "Functional Walk Tests in Individuals with Stroke: Relation to Perceived Exertion and Myocardial Exertion." *Stroke* 33 (3): 756–761.

Engman, Athena. 2019 "Embodiment and the Foundation of Biographical Disruption." *Social Science & Medicine* 225: 120–127.

Faircloth, Christopher A., Craig Boylstein, Maude Rittman, and Jaber F. Gubrium. 2005. "Constructing the Stroke: Sudden-Onset Narratives of Stroke Survivors." *Qualitative Health Research* 15 (7): 928–941.

Faircloth, Christopher A., Craig Boylstein, Maude Rittman, Mary Ellen Young, and Jaber Gubrium. 2004. "Disrupted Bodies: Experiencing the Newly Limited Body in Stroke." *Symbolic Interaction* 27 (1): 71–87.

Gadow, Sally. 1980. "Body and Self: A Dialectic." *The Journal of Medicine and Philosophy* 5 (3): 172–185.

Giddens, Anthony. 1991. *Modernity and Self-identity: Self and Society in the Late Modern Age*. Stanford, CA: Stanford University Press.

Goggin, Gerard. 2016 "Disability and Mobilities: Evening Up Social Futures." *Mobilities* 11 (4): 533–541.

Gyllensten, Amanda Lundvik, Lisa Skär, Michael Miller, and Gunvor Gard. 2010. "Embodied Identity—A Deeper Understanding of Body Awareness." *Physiotherapy Theory and Practice* 26 (7): 439–446.

Head, Henry, and Gordon Holmes. 1911. "Sensory Disturbances from Cerebral Lesions." *Brain* 34 (2–3): 102–254.

Imaizumi, Shu, Tomohisa Asai, and Shinichi Koyama. 2016. "Embodied Prosthetic Arm Stabilizes Body Posture, While Unembodied One Perturbs It." *Consciousness and Cognition* 45: 75–88.

Janssens, Maria Catherina Wilson. 2018. "Spatial Mobility and Social Becoming: The Journeys of Four Central African Students in Congo-Kinshasa." *Geoforum*. https://doi.org/10.1016/j.geoforum.2018.05.018.

Kaufmann, Vincent, Manfred Max Bergman, and Dominique Joye. 2004. "Motility: Mobility as Capital." *International Journal of Urban and Regional Research* 28 (4): 745–756.

Kirkevold, Marit, Doris Christensen, Grethe Andersen, Søren Paaske Johansen, and Ingegerd Harder. 2012. "Fatigue After Stroke: Manifestations and Strategies." *Disability and Rehabilitation* 34 (8): 665–670.

Kitchin, Rob, and Robin Law. 2001. "The Socio-Spatial Construction of (In)Accessible Public Toilets." *Urban Studies* 38 (2): 287–298.

Kitzmüller, Gabriele, Terttu Häggström, and Kenneth Asplund. 2013. "Living an Unfamiliar Body: The Significance of the Long-Term Influence of Bodily Changes on the Perception of Self After Stroke." *Medicine, Health Care and Philosophy* 16 (1): 19–29.

Klinke, Marianne E., Dan Zahavi, Haukur Hjaltason, Björn Thorsteinsson, and Helga Jónsdóttir. 2015. "'Getting the Left Right': The Experience of Hemispatial Neglect After Stroke." *Qualitative Health Research* 25 (12): 1623–1636.

Kristensen, Hanne Kaae, Malin Tistad, Lena Von Koch, and Charlotte Ytterberg. 2016. "The Importance of Patient Involvement in Stroke Rehabilitation." *PloS One* 11 (6): e0157149. https://doi.org/10.1371/journal.pone.0157149.

Kvigne, Kari, and Marit Kirkevold. 2003. "Living with Bodily Strangeness: Women's Experiences of Their Changing and Unpredictable Body Following a Stroke." *Qualitative Health Research* 13 (9): 1291–1310.

la Cour, Peter, and Marian Petersen. 2015. "Effects of Mindfulness Meditation on Chronic Pain: A Randomized Controlled Trial." *Pain Medicine* 16 (4): 641–652.

Lawrence, Enas S., Catherine Coshall, Ruth Dundas, Judy Stewart, Anthony G. Rudd, Robin Howard, and Charles DA Wolfe. 2001. "Estimates of the Prevalence of Acute Stroke Impairments and Disability in a Multiethnic Population." *Stroke* 32 (6) (2001): 1279–1284.

Leder, Drew. 1990. *The Absent Body*. University of Chicago Press.

Loja, Ema, Maria Emília Costa, Bill Hughes, and Isabel Menezes. 2013. "Disability, Embodiment and Ableism: Stories of Resistance." *Disability & Society* 28 (2): 190–203.

Merleau-Ponty, Maurice. 2001. *Phenomenology of Perception*. Translated by C. Smith. London: Routledge.

Mitchell, Sue. 2018 "Open Wounds: Visible Physical Disability and Its Meaning for the Group and Society." *Group Analysis* 51 (4): 487–499.

Murray, Stuart J., and Dave Holmes. 2014. "Interpretive Phenomenological Analysis (IPA) and the Ethics of Body and Place: Critical Methodological Reflections." *Human Studies* 37 (1): 15–30.

Nijboer, Tanja, Ingrid Van de Port, Vera Schepers, Marcel Post, and Anne Visser-Meily. 2013. "Predicting Functional Outcome After Stroke: The Influence of Neglect on Basic Activities in Daily Living." *Frontiers in Human Neuroscience* 7: 182. https://doi.org/10.3389/fnhum.2013.00182.

Norlyk, Annelise, Bente Martinsen, and Klaus Kjaer-Petersen. 2013. "Living with Clipped Wings—Patients' Experience of Losing a Leg." *International Journal of Qualitative Studies on Health and Well-Being* 8 (1): 21891. https://doi.org/10.3402/qhw.v8i0.21891.

Papadimitriou, Christina. 2008. "Becoming En-Wheeled: The Situated Accomplishment of Re-Embodiment as a Wheelchair User After Spinal Cord Injury." *Disability & Society* 23 (7): 691–704.

Pettit, Harry, and Wiebe Ruijtenberg. 2019. "Migration as Hope and Depression: Existential Im/Mobilities in and Beyond Egypt." *Mobilities*, 1–15.

Pretorius, Chrisma, and Ninon Joubert. 2014. "The Experiences of Individuals with Multiple Sclerosis in the Western Cape, South Africa." *Health SA Gesondheid* 19 (1). https://doi.org/10.4102/hsag.v19i1.756.

Rice, Carla, Eliza Chandler, Jen Rinaldi, Nadine Changfoot, Kirsty Liddiard, Roxanne Mykitiuk, and Ingrid Mündel. 2017. "Imagining Disability Futurities." *Hypatia* 32 (2): 213–229.

Rubio, Kerry Brockmann, and Julia Van Deusen. 1995. "Relation of Perceptual and Body Image Dysfunction to Activities of Daily Living of Persons After Stroke." *The American Journal of Occupational Therapy* 49 (6): 551–559.

Sakellariou, Dikaios, Gail Boniface, and Paul Brown. 2013. "Experiences of Living with Motor Neurone Disease: A Review of Qualitative Research." *Disability and Rehabilitation* 35 (21): 1765–1773. https://doi.org/10.3109/09638288.2012.753118.

Sakson-Obada, Olga, and Jowita Wycisk. 2015. "The Body Self and the Frequency, Intensity and Acceptance of Menopausal Symptoms." *Menopause Review* 14 (2): 82.

Salazar, Noel B. 2012. "Tourism Imaginaries: A Conceptual Approach." *Annals of Tourism Research* 39 (2): 863–882.

Salazar, Noel B., and Alan Smart. 2011. "Anthropological Takes on (Im)Mobility." *Identities* 18 (6): i–ix.

Schuster, Corina, Andrea Glässel, Anne Scheidhauer, Thierry Ettlin, and Jenny Butler. 2012. "Motor Imagery Experiences and Use: Asking Patients After Stroke Where, When, What, Why, and How They Use Imagery: A Qualitative Investigation." *Stroke Research and Treatment*. https://doi.org/10.1155/2012/503190.

Scott-Sheldon, Lori A. J., Emily C. Gathright, Marissa L. Donahue, Brittany Balletto, Melissa M. Feulner, Julie DeCosta, Dean G. Cruess, Rena R. Wing, Michael P. Carey, and Elena Salmoirago-Blotcher. 2019. "Mindfulness-Based Interventions for Adults with Cardiovascular Disease: A Systematic Review and Meta-Analysis." *Annals of Behavioral Medicine*. https://doi.org/10.1093/abm/kaz020.

Smith, Laura J., and Rachel L. Shaw. 2017. "Learning to Live with Parkinson's Disease in the Family Unit: An Interpretative Phenomenological Analysis of Well-Being." *Medicine, Health Care and Philosophy* 20 (1): 13–21. https://doi.org/10.1007/s11019-016-9716-3.

Stott, Hannah. 2019. "Embodiment, Altered Perception and Comfort After Stroke." PhD diss., University of the West of England.

Tham, Kerstin, Lena Borell, and Anders Gustavsson. 2000. "The Discovery of Disability: A Phenomenological Study of Unilateral Neglect." *The American Journal of Occupational Therapy* 54 (4): 398–406.

Townend, Ellen, Deborah Tinson, Joseph Kwan, and Michael Sharpe. 2010. "'Feeling Sad and Useless': An Investigation into Personal Acceptance of Disability and Its Association with Depression Following Stroke." *Clinical Rehabilitation* 24 (6): 555–564.

Tyson, Sarah F., Marie Hanley, Jay Chillala, Andrea B. Selley, and Raymond C. Tallis. 2008. "Sensory Loss in Hospital-Admitted People with Stroke: Characteristics, Associated Factors, and Relationship with Function." *Neurorehabilitation and Neural Repair* 22 (2): 166–172.

van der Meide, Hanneke, Truus Teunissen, Pascal Collard, Merel Visse, Leo H. Visser. 2018. "The Mindful Body: A Phenomenology of the Body with Multiple Sclerosis." *Qualitative Health Research* 28 (14): 2239–2249.

Veal, Charlotte. 2017. "Micro-Bodily Mobilities: Choreographing Geographies and Mobilities of Dance and Disability." *Area* 50 (3): 306–313.

Vigh, Henrik. 2006. "Social Death and Violent Life Chances." In *Navigating Youth, Generating Adulthood: Social Being in an African Context*, edited by Catrine Christiansen, Mats Utas, and Henrik Vigh. Nordiska Afrikainstitutet, 31–60.

Vigh, Henrik. 2009. "Motion Squared: A Second Look at the Concept of Social Navigation." *Anthropological Theory* 9 (4): 419–438.

9

"When You Do Nothing You Die a Little Bit": On Stillness and Honing Responsive Existence Among Community-Dwelling People with Dementia

Laura H. Vermeulen

Introduction

"She might be downstairs, drinking coffee. She's always on the move!" The care worker, looking up from her files, smiled at me. I had popped by the nurses' station to inquire about 91-year-old Mrs Kelk. Mrs Kelk and I ran into each other downstairs at the point where the reception desk, the elevators and the entrance met. A small figure, Mrs Kelk's fingers were searching for a button on the elevator panel. "I know who you are!" she said laughing when she saw me, mimicking my playful way of trying to remember who she was, and continued: "I'd like to… I thought I'd go for a little…" She closed her eyes, frowned and nodded, then swiftly turned her walker towards the entrance: "Are you coming along?!" A moment later, we found ourselves sitting on a bench at the entrance. She started telling me about the coffee room she had just left, pointing in its direction: "I thought I'd go for a little walk. If I sit there

---

L. H. Vermeulen (✉)
University of Amsterdam, Amsterdam, The Netherlands
e-mail: l.h.vermeulen@uva.nl

nothing happens. Then you see nothing. You want to do something, but you don't know how. Then you do nothing. And then you die". "You die?!" I asked her. She looked up at me and nodded. "You don't really die, but a little bit. There is no music in it!" Mrs Kelk was silent for a while, then she continued to tell me that there was no real point to life anymore, but that one had to make something of it oneself. "So then I leave", she said, "like that...", and she turned her head towards the glass door which looked out onto the busy road bordering the care home.

Listening to Mrs Kelk, I was struck by the way in which she associated sitting in the coffee room, where nothing happened, with death and non-involvement, and how she used this association between immobility, non-involvement and death to explain why she had wanted to go for a walk that morning, as well as on other occasions. From the moments we spent in the coffee room together later that year, I surmised that she could have sat at a table with others and that her impaired hearing prevented her from engaging in group conversations. Hence, I thought that her complaint could have been directed towards a lack of meaningful social interaction. Yet following her descriptions of the situation, it seemed to be about something more encompassing: about nothing happening at all, about wanting to do something but not knowing how, about not seeing anything. While she did nuance her dramatic capturing of this moment in terms of "dying" by emphasizing that she was using the word metaphorically, there was something about her way of handling it—going for a walk—which suggested that she felt the impact of this moment viscerally, as something that had to be acted on. Her ensuing "so then I leave" complicated this reading, however. It suggested that going out was more than a pre-reflexive response: a strategy that helped her to deal with the estranging moment. Puzzled by her way of handling the still scene, I turned to the accounts of my other interlocutors. Mrs Kelk was particularly clear in explaining the association between immobility and non-involvement and the need to leave. But she was not alone in making this connection.

This chapter investigates the recursive mobilities of leaving and returning home that community-dwelling, single-living people with dementia in one of the larger cities in the Netherlands engaged in. It traces these movements to moments that my interlocutors described as "still". Joining recent contributions to critical phenomenology (Dyring and Grøn, n.d.; Zigon 2018; Guenther 2013) and drawing on an anthropology of responsiveness (Leistle 2017; Louw 2017; Waldenfels 2011), I conceptualize these still moments as "potentiating intervals" (Dyring and Grøn, n.d.) that sparked recursive practices of crafting the very responsive engagement with the world that life exists in. In doing so, I aim to contribute to studies that reconsider stillness and waiting as phenomena that trigger creativity (Janeja and Bandak 2018; Ehn and Löfgren 2010, 165; Bissell 2008). In pursuing this line of thinking, I do not only respond to calls for a social model that enables us to move beyond explanations that pathologize mobility in people with dementia (McDuff and Phinney 2015; Dewing 2006; see also Van Wijngaarden et al. 2019; The 2017). I also add to arguments that underline potentiality in lives lived with a condition that is otherwise considered to foreclose meaningful futures (Taylor 2017; see also Higgs and Gilleard 2015). I maintain that by attending closely to how repetitive mobilities of leaving and returning home could be seen as attempts to nurture capacities to responsively engage the world, we may draw broader lessons on how we all cope with vulnerable moments in our lives. Understood in this way, life and caregiving may be reconceptualized as honing our abilities to respond (see also Leistle 2017, 40).

Stillness in the Lives of People with Dementia

I met Mrs Kelk during a 2-year period of ethnographic research on the ways in which people with dementia, living in the community in single-person households, upheld their daily lives. Like my other interlocutors, Mrs Kelk was diagnosed with (early stage) dementia, a syndrome that medical doctors define as a "syndrome of cognitive impairment that affects memory, cognitive abilities and behaviour, and significantly interferes with a person's ability to perform daily activities" (WHO 2018, 6). Mrs Kelk differed from my other interlocutors—all of whom were 67 years of age and older, living in the community on their own, and receiving occasional help from home care services—because of her old age and the fact that she was living in an elderly care institution which offered her an apartment of her own and 24-7 care. Yet her account of daily life was similar to the others'. Struck by the way in which front doors appeared as major obstacles in my research collaborators' daily lives—keys got lost and doors were often returned to for a check of whether they were locked—I began to discuss with them their frequent movements of leaving and returning home. In all of their stories, I found metaphors of death and non-involvement associated with staying inside and doing nothing on the one hand and a stress on "going out" on the other. I grouped together as "stillness" the moments that they described as "staying put", the situations they associated with death and dying, with not "seeing or hearing anything" and with "not knowing what to do".

In this chapter, I take the vicissitudes of these still moments as a starting point to investigate why going out was so important to my interlocutors living with dementia. Following Hage's (2009) approach of moving beyond metaphor to see how moments of (im)mobility are imagined/felt, I trace how in still moments, involvement was interrupted on the levels of practical engagement, social relations and existence. I remain close to my initial understanding of stillness as pertaining to moments when the ability to engage life receded from grasp. I do so by investigating how the experiences that my interlocutors referred to with metaphors of death and deterioration sprang from the property that life assumed for them in "moments in-between actions where purpose and direction ebb" (Harrison 2011, 209). While some scholars have framed still moments in positive terms—as moments of calm (Conradson 2011; Stewart 2007) or of demonstration in a political sense (Harrison 2011)— my interest is in what Bissell and Fuller have termed stillness' "capacity to garner suspicion" (2011, 7), its ability to confound and unsettle commonplace modes of comprehension.

I wish to illustrate how, for my interlocutors, stillness intervened in the relational quality of life. It did so by disturbing everyday engagements (not knowing what to do or how to do things), interrupting felt relations (neither seeing nor hearing anyone) and unhinging existential relationality (questioning who one is) (Guenther 2013, 214). I show how walking was my interlocutors' creative response to such an interruption, a response that allowed new ways of inhabiting the world to arise (Dyring and Grøn, n.d.). In doing so, I join recent contributions in critical phenomenology that assume lived experience to be fundamentally relational (Guenther 2013), and which study these moments for the potential they harbour to disclose new possibilities (Dyring and Grøn, n.d.; Zigon 2018, 17). In particular, I draw on the work of those who write about the generative character of experiences of estrangement (Dyring and Grøn, n.d.; Waldenfels 2011; Wentzer 2018; Leistle 2017; Louw 2017). I maintain that this approach is helpful for our purposes at hand, first of all, because reformulating existence as a responsive process (Waldenfels 2011, 36; see also Zigon 2018, 171) allows us to think of mobility and immobility together as iterations of the same dialogical process. Secondly, instead of studying people diagnosed with dementia as "other", it enables us to join them in thinking about our own responses to predicaments of life that we all may face (see Leistle 2017, 8–9).

A Case of Wandering? De-medicalizing Dementia

My differences, seen in how I act, should not be seen as abnormal and unnatural but just as a sort of a diversity of what is possible in all people. As my dementia creeps into me, I am aware that it gets harder and harder to be just different and not labelled or rated on a test. Things I did before, now have become dementia-related behaviours. I am wandering now, and what scares me is not wandering. I am open to this and what it might offer me, but that others will see this as bad and take it on as a problem for them (Lady living with dementia, quoted in: Dewing 2006, 242).

Living with dementia is associated with both fine motor movements (movements of the hands) and repetitive mobilities (frequent walking). It is considered a major cause of disability and dependency among a quarter of all adults aged 80 years and over (Alzheimer Nederland 2019). With no cure or effective prevention within reach, managing (im)mobility in people with dementia is among the key focuses of dementia strategies across the world. Extensive, unsupervised movement in people living with the condition is described as a key safety and health threat (Yuchi Young et al. 2018; Cipriani et al. 2014; Matteson and Linton 1996 in Joller et al. 2013). Coined "wandering", it is considered a form of agitated behaviour particular to dementia that is particularly challenging to carers (Hope et al. 2001 in Joller et al. 2013), and it constitutes a major reason for nursing home admission (Cipriani et al. 2014).

As the woman quoted at the opening of this section suggests, however, considering mobility as a problem related to dementia is not self-evident to all. There is growing interest in the cultural underpinnings of the labelling of mobility in people with dementia as either meaningful or meaningless (Solomon and Lawlor 2018; Brittain et al. 2017; Dewing 2006) and in practices that allow for positive attachments to the care home (Driessen et al. 2018). Yet, despite calls for studies offering alternative explanations of wandering by drawing on the experiences of people engaging in mobility themselves (Dewing 2006, 247), with the exception of one study I know of (McDuff and Phinney 2015), little attention has been paid to experiences that inform (im)mobility in people living with dementia. What gives rise to the urge among persons with dementia to (repeatedly) leave the house? How do these experiences shape the mobilities arising from them? And what, in turn, do such mobilities offer to the experiences they originate from?

Seeking answers to these questions, I argue, allows us to highlight how "corporeal existence is composed through vulnerability and withdrawal as much as through action and worldly engagement" (Harrison 2008, 425). Answering these questions also helps us to respond to calls for case studies of mobility among people with dementia that go beyond medicalization (McDuff and Phinney 2015; Dewing 2006), and which encourage us to go beyond arguments that show how wandering could be something natural from the perspective of people with dementia themselves (Dewing 2006, 247). That is, it may inspire us to take what people with dementia tell us not as cases of living with a disease, but instead as cases of life. As the lady interviewed by Jan Dewing, who was quoted at the start of this section, put it, the experiences of people with dementia could then become "a diversity of what is possible in all people" (2006, 242).

In the following, I draw on the accounts of people impacted to varying extents by a loss of memory and orientation. While I do foreground the conversational exchanges that made most sense to me, I do not wish to qualify the statements used here as more "lucid" or "valid" than other communication threads that I was less equipped to deal with. Nor do I wish to rate the validity of the different accounts used in this chapter. In line with the social model outlined here, I maintain that working with people with cognitive losses requires interpretation work that is different from the interpretation involved in working with those without cognitive losses, but that this difference is not one in kind but one in degree.

Mrs Kelk and the Under-Arousal Hypothesis

The tension between medical and social approaches to Mrs Kelk's habit of going for a walk was the focus of discussions about her residence in the elderly care home between Mrs Kelk's children and the nursing staff, between September 2014 and September 2015. The institution where Mrs Kelk lived was one of the few that had survived the budget cuts and discourses about ageing-in-place that had led to either the closure of other nursing homes or their up-scaling to nursing homes with closed-door policies for people with dementia. Mrs Kelk was therefore one among just a few people with dementia in the Netherlands who lived in an environment which offered her an apartment of her own, 24-7 care, and which also allowed her to move around freely. This latter aspect is where tensions arose.

I had come to know Mrs Kelk as an independent widow, a mother of four and the second daughter in a family of nine. She had lost two brothers who were engaged in an anti-Nazi revolt and spent her later life fighting for the recognition of their work. She had travelled to Cuba to support the socialist cause, was wary of anything kitsch, prided herself on being helpful to others, yet hated people wanting to be liked. Furthermore, going for a walk was one of the things she both liked and did frequently; indeed, my own presence in Mrs Kelk's life was legitimized by my ability to support her in going out more often. Mrs Kelk's children encouraged her to go out often, as it provided her with something positive to do during the day. They also accepted the risk of her having an accident on the busy road bordering the home, considering that the risk of diminishing her quality of life by restraining her from going out would be far greater. Yet care workers dreaded the idea of anything happening to Mrs Kelk during their shift. Given that Mrs Kelk herself expressed anger at being constrained from going out, her walks gave rise to two consultations that brought together Mrs Kelk's daughter Alex, her sons Kees and Jack, care worker Nathalie, locality manager Gerda and geriatrician Nienke (all pseudonyms).

In the first consultation in September 2014, Mrs Kelk's walks were explained in terms of her joy in getting fresh air (Jack and Nathalie), by the habit she had established when rehabilitating from an operation on her shoulder (Kees), and by her increasing feeling of not knowing what to do and no longer recognizing her own apartment (Alex). At the time of the second meeting a year later, in September 2015, things seemed to have changed. Mrs Kelk had difficulties finding her way out of the building and she could be observed wandering the corridors of the care home for hours on end. Given that this "extensive wandering" put her, according to the carers, at risk of falling and experiencing agitation, the consultation centred on how to calm her. In the first consultation, when talking about her walks, Nienke had built on psychosocial models heralding Mrs Kelk's identity as an active and independent woman. Yet in this second meeting, she started off by saying that she had recently had a conversation with the neurological geriatrician Anneke van der Plaats, whose approach she considered a suitable tool for "explaining behaviour in advanced dementia". She therefore wanted to share some insights along Van der Plaats' line of reasoning.

The "problem" for Mrs Kelk, according to Nienke, was an inner unrest, which she thought became uncontrollable for her in moments when nothing was happening. These were moments in which Mrs Kelk "did not know what to do", Nienke thought. She would then start walking and the walking would become incessant. Talking about the phenomenon manifesting at night, Nienke explained:

> It could be the case that it is too silent, that she has too little arousal, and that because of that it feels as though she enters the infinite. She will start seeking, and what might be of help is a clock ticking or another kind of noise. For this was the suggestion from the conversation [with Anneke van der Plaats], to bring in a motile form of arousal. Because

things which don't move become part of a background falling away. So there has to be something moving. And that if you wake up at night, that you…

Alex: … there is something comforting.

Nienke: … well, that there is something catching attention, giving comfort and that there is not… nothing.

Nienke linked silence at night with the phenomenon of under-arousal, a situation in which "everything is static and nothing happens" (Van der Plaats and De Boer 2014, 18). In mentioning the "infinite", she also built on Van der Plaats' neurology-inspired work to stress the importance of there being "something" as opposed to "nothing". Later in the meeting, she explained that people with dementia could potentially lose the capacity to process static forms of arousal and that, as a consequence, they could stop feeling alive in situations lacking dynamic arousal (ibid., 22). This feeling she compared to the lack of feeling one has in a leg that one does not stretch for a while.

Nienke's description of Mrs Kelk's discomfort at nothing happening resonated with most of us who knew that Mrs Kelk more often compared doing nothing with dying. Yet, even though Nienke's solutions based on Van der Plaats' model of thinking were briefly discussed in the meeting, they were soon discarded. Mrs Kelk's hearing was deemed too impaired to use background noise as a form of dynamic arousal without disturbing her neighbours, and bringing moving lights into the room was deemed a potential source of unrest in itself, since lights lacking any obvious purpose would, in Mrs Kelk's opinion, cost money and would have to be switched off. Instead, another of Nienke's ideas was agreed on: care workers would be asked to sit with Mrs Kelk for twenty minutes each morning, talking with her attentively so as to "really reach" and comfort her.

This solution proved to be hard to fit into the busy schedules of the care workers, and I found myself accompanying Mrs Kelk on hour-long walks in the autumn of 2015. During these walks—we carefully looked in many corners and waited at numerous doors—Mrs Kelk occasionally told me that she was looking for "a group of people" whom she could join or for one of her children who could tell her things so that she "would know stuff again at least". I wondered whether in the consultations we had really understood how Mrs Kelk seemed to be looking for satisfying forms of engagement with the world around her. I started to compare her walks with the frequent tours around the house of my other interlocutors living with dementia.

Stillness of Life

When they described what they were bothered by when at home alone, my interlocutors with dementia used the adjectives "dead", "deafening" and "very silent". They also told me that the silence pervading their homes could be as "quiet as a mouse". In saying this, they were talking about more than the absence of sound. Stillness itself felt like a lack of involvement of the senses. "It is more like living in a void now. I see nobody, I hear nobody, nobody comes, nobody phones", Mrs Twijnders never failed to tell me when I called her and asked her, for an opening, how she was doing. For 74-year-old Willem, the built environment was involved. When I asked him what he thought of our new habit of meeting up and talking about his life, he would tell me that "the walls [were] chatting again". There were not many people coming up to his apartment these days. Els, his daughter, did come occasionally. But the butcher and the baker? They did not. Seventy-nine-year-old Mrs Huisman was known for advising people that she was never to be found at home after 2 p.m. "People die at home", she used to say, explaining why, over the last couple of years, she had never let a day pass without taking a trip to the shopping mall.

The stillness pervading their lives was social in character. With the situation of family members passing away or becoming ill or absorbed with caring, with friends changing political convictions and losing contact, or they themselves losing their capacities to host friends for dinner, the established networks of my respondents were shrinking. Whereas some would link the advance of this stillness to the death of a loved one, their dementia was also involved in different ways. While Rob told me that acquaintances had backed away from talking to him upon hearing about his diagnosis, Mrs Twijnders admitted that forgetting about her daughter's visit a few days ago played a role in her experience of being just by herself.

Stillness also seemed to manifest itself in my interlocutors' practical engagements. "People die at home", was Mrs Huisman's shorthand way of explaining her routine of going out every afternoon. Like many other older adults I met, she could tell me that engaging in physical exercise was a means of maintaining better health in general. But when I asked Mrs Huisman what exactly she did not like about being at home in the afternoon, she explained that in the afternoon she would find herself "in the boat of 'What am I doing here?'"—and I am here mimicking her incomplete use of the well-known Dutch expression "being in the same boat". In the morning, there were tasks to be done around the house, Mrs Huisman said, but in the afternoon, these tasks seemed to have disappeared.

Experiences such as that alluded to by Mrs Huisman sprung from the fact that old routines were in disarray. Many of my interlocutors had lived out Dutch post-war working-class routines stipulating that married women took care of housework in the morning and looked after the children and grocery shopping in the afternoon, while men went out to do maintenance work or paid labour during the day. And yet there seemed to be more at work here. In repeatedly using metaphors depicting a loss of place, many of my interlocutors told me that their way of coping in the world had become hampered; such as when Willem, in the middle of figuring out where the onions were supposed to go, exclaimed "I lost my house!", before going on to add, "It is the automatism going missing"; and as when Rob fled his apartment, frustrated at all the maintenance work that needed to be done and that now took such an incredible amount of effort and time. In some cases, activities themselves became obtrusive, such as when people explained that they no longer read books like they used to. In other cases, activities of daily living simply disappeared from sight, such as when Mrs Kelk asked her daughter what she was actually supposed to do in her apartment.

The various aspects making up stillness might have derived their uncanniness from the fact that they were weaker forms of a process that was uprooting existence in a more fundamental manner. I understood this as happening one evening during a conversation I had with Mrs

Kelk, not long after the multidisciplinary consultation in 2015. Mrs Kelk had become angry at the fact that she knew so little about a daycentre I had mentioned. She had stopped responding to my questions and had gradually sunk back into her own thoughts. Wringing her hands, her eyes started searching, and her voice became rasping and turned to a high pitch as she asked: "So, who then, am I...? [...] I don't know that. [...] I have nothing... I don't know... that". After that, her eyes fixed on the door, she got to her feet in a painstakingly slow manner. She shuffled past me, opened the front door, walked through it and turned left, disappearing from sight.

The Generative Potential of the Liminal

In writing about stillness, others have rendered it moments of "stasis without projection and orientation" (Harrison 2011, 9). In stillness, purpose and direction ebb, there is an absence of action and signification (ibid., 210). According to Bissell and Fuller, this absence may become dynamic. "If doings, actions and practices are the source and locus of signification", they write, "the still is confounding and unsettles common place modes of comprehension. It might denote a radical withdrawal. [As such, it] has the capacity to garner suspicion" (2011, 7). The different iterations of stillness described above might have been illustrative of this. They spoke of situations in which life assumed the character of a vacuum (see also Ehn and Löfgren 2010, 22; Bissell 2008). The self-evident intelligibility of my interlocutors involved being-in-the-world became obtrusive on the level of personal relations, social positionings, physical engagement and existential involvement. My evening conversation with Mrs Kelk may be an example of this: in that moment, the very intelligibility of her involved being-in-the-world seemed to have withdrawn from her grasp.

In order to explore the generative quality of these estranging, still moments, I draw on the work of Bernhard Waldenfels (2011) and others who build on his writings (Leistle 2017; Wentzer 2018; Louw 2017; Dyring and Grøn, n.d.; Dyring 2015). Waldenfels considers disturbances to familiarity to be a key dynamic of human life. Starting from the writings of Husserl, Merleau-Ponty and Levinas, he conceives of existence as a responsive process (2011, 36; Leistle 2017, 260; Zigon 2018, 171). Human beings, according to him, are essentially liminal beings, in that they find themselves in a process of responding to a field of indeterminacy that calls for their response. This field of indeterminacy Waldenfels calls "the alien" (2011, 10–11). Essentially ambiguous, the alien is "something that cannot be pinned down" (ibid., 3). "Before it arises as a theme, it makes itself know as disruption, interference or disturbance […] that interrupts familiar formations of sense and rule" (ibid., 81; 36). Human beings' implicit responsive envelopment with alterity, according to Waldenfels, plays out constantly and on different levels. It manifests itself in everyday interactions, such as in the encounter with an unknown other (Leistle 2014, 66). It appeals to us in our confrontation with practices belonging to an order we consider different from our own, such as a ritual or a language we do not understand (ibid.). Lastly, it makes itself known in radical alienness. While we encounter alienness in this latter, radical sense to some degree in all spheres of life, it is most explicitly experienced in phenomena such as birth, death, pain, illness, shock, trauma and dreams (Leistle 2017, 37). In these phenomena, "we find not only our interpretations questioned but also the possibility of interpreting itself" (Leistle 2014, 66; see also Louw 2017). Key to this discussion is the generative potential inherent in the responsive dynamic that Waldenfels describes. Given the illusive character of alterity, Waldenfels argues, a difference remains between alterity's original appeal and the concrete answer we come up with when responding to it. That is, there is an interval or responsive difference between the "what" of the response and the "to what" of responding. The fact that this interval can never be closed (Waldenfels 2011, 5; 31), makes for both the urgency of responding—addressed as someone "able-to" respond (Dyring 2018), we cannot "not" respond (Dyring et al. 2018, 27)—and for the creative potential in all experience. That is, any answer we come up with is partial, and could hence be infinitely multiple, in the face of the excessive and ungraspable character of the original appeal.

In an upcoming contribution to a critical phenomenology of potentiality and life with dementia, Dyring and Grøn (n.d.) elaborate on the creative potential engendered by experiences of alterity. They draw on Waldenfels' conceptions of interruption and the interval to argue how ambiguous experiences could open new registers of possibility (ibid., 12). Alterity, they write, can be explored phenomenologically "in the more or less obtrusive interruptions that occur between people in everyday life—often in ways that might be disturbing and uncomfortable, yet also ethically creative and generative of meaning" (ibid., 7). Such interruptions are interruptions in a literal sense, they write. They are "eruptions of potentiated intervals, making room for new ways of responding, and of joining together in responding, to the challenges and prospects these interruptions put to those affected by them" (ibid., 7–8).

Dyring and Grøn make their argument following Else, a lady with dementia, in her first months on the dementia ward. When Lone, the second author, first met Else, she was walking the hallway day and night. With her bags packed, and wearing an overcoat and scarf, she would pull at doors and ask whether she was in a prison. In this situation, the Little One appeared: a stuffed toy animal that Else said she had found abandoned. Over the months, Else built an affective and caring relationship to the indeterminate being that alternated from being a baby, a cat and a toy to her. Following these events, the authors write of Else as being "interrupted" in her ways of inhabiting the ward by the encounter with the ambiguous figure of the Little One. In this interruption, they argue, "a potentiated interval erupts […] allowing for another mode of inhabiting the ward" (ibid., 26). This new way of inhabiting the world is different from the fraught situations in which Else found herself when she first came to the ward. It is a situation punctuated with meaning and a sense of purpose that did not seem to be there prior to the Little One's arrival (ibid.).

In the lives of my interlocutors, we might see a similar role played by stillness, as both an uncanny interruption to familiarity and an anarchic element generative of new ways of inhabiting the world. Their use of metaphors of death, dying and their equivalents of loss may indicate how moments of stillness appeared to them as a more or less unwelcome disturbance to their way of being comfortably in the world with themselves and others. In its most profound sense, as witnessed during my evening conversation with Mrs Kelk, the withdrawal of the possibility of interpretation as such might have been at stake. Yet, however, unsettling these estranging, still moments were, they also seemed to open up new possibilities.

In the following, I will substantiate how this could work. I do so by giving a detailed account of how stillness could propel engagement for my interlocutors. I follow stillness' generative quality, on both an everyday and an existential level. Drawing on the story of one of my interlocutors in particular, I argue how going out might have been the response enabled in the interval opened up by stillness; and how, in going out, my interlocutors may have crafted ways of inhabiting the world comfortably anew, ways of responsively being-in-the-world that still life could not engage them in.

Breaking the Silence by Going Out

When asked what they did when they were bothered by stillness at home, many interlocutors explained that they left the house and sought different forms of engagement. Rob said that he "fled" his maintenance work, going for a coffee and a free newspaper outside. Liz told me that she "escaped" her sense of not knowing what to do by going for a walk and talking to people out with their dogs. Wieke simply left her house, sometimes up to 18 times a day. And unlike Mrs Twijnders, who roamed the suburbs reminiscing about who used to live there when she was young, Mrs Huisman sought real-life encounters: she always headed for the shopping mall at 2 p.m., where she would walk about, talk to shopkeepers and sit on benches, chatting with other elderly ladies.

All these practices seemed helpful to my interlocutors, because they enabled them to relate anew, reaching beyond broken routines and loved ones lost. When returning home in the evenings, Mrs Huisman could tell me that she had spent her day very sociably. She had seen different things, and she could tell me lots of stories about the many people she had met. Yet in structuring days with little trips outside, another more fundamental process also seemed to be secured. Liz told me that the situation she found herself in was very different when she returned home after her walks around the neighbourhood; whereas she fled the house not knowing what to do, upon entering the front door, she would look around and tell herself: "Well I could start doing this, and doing that…".

It was 84-year-old Jan who could explain how going out sparked ways of engaging life at home anew. One day, in a conversation we had about the "silence of the walls" that others had told me about, Jan agreed that the house indeed could be "dead silent". He explained how in such very quiet moments, he would find himself sitting on the couch with nothing particular to do. He told me that he could start "messing around", but that he did not feel like it. And he could "break the silence" by making some noise himself, he said, "but this is not what you want to hear […] You're listening because you're listening to something". He went on:

J: And then I mostly tend to go for a little walk.
L: Yes?
J: And… break that silence.
L: Really? How do you break the silence when going for a walk?
J: You're starting to think what you would want to say…. But you don't have anyone to say it to, so… you don't, but, but in your mind you have got the thoughts from that.
L: Thoughts from that… okay…?
J: Yes, what you would want to say! If you were to run into somebody, you'd say [mimicking a high, loud tone of voice]: "Heeeeey! How are you?!" Well, that person could just break… that loneliness….
L: Yeah, I see. And that person does so by…?
J: Well, by saying: "How are you doing?" […]
L: [And] what happens to the loneliness when you're dealing with it in this way?
J: Well, […] then it's not lone – lonely anymore. You have people around you. It's true they say nothing…!

At this moment in the conversation, tellingly, Jan's phone rang. But the fragment reveals how the impact of silence seemed both to viscerally interrupt Jan's way of comfortably being-in-the-world and to urge him to go out. In silent moments, modes of engaging that were otherwise available to Jan—messing around and making noise himself—withdrew from his grasp. And not unlike the way in which Mrs Kelk characterized sitting in the coffee room without seeing or doing anything as dying a little bit, Jan described how, in still moments, a more fundamental form of engagement—the (imagined) or real-life engagement with other people that most of us for most of our lives dwell in—was disturbed. The interval opened up by stillness did more than simply urge him to take up another task. Going for a walk, at the risk of meeting other people, helped Jan, because it involved him in a responsive engagement with the world again.

Honing Responsive Engagement as a Way to Live

Elsewhere it is argued how various forms of mobility may spill over into each other. The way in which walking engaged Jan in an imagined dialogue with others may have been well explained by Hage who suggests that existential and physical movement are intertwined (2005), and by Ehn and Löfgren who illustrate that walking could be a prerequisite for thinking and daydreaming (2010, 139). In a similar vein, Dyring and Grøn show how in Else's caring for the toy animal, a new, caring, dimension opened up in her way of inhabiting the ward (n.d., 26). In the case of the disturbing moments marked by metaphors of death in this chapter, the estranging impact of stillness might have opened up room for forms of relating that allowed my interlocutors to engage the world as a familiar place again. When physically going out, Jan started to plot imagined encounters and conversations. Doing so he might have honed the responsive engagement—the ability to viscerally and dialogically engage with oneself, the world and others—that had been stilled in moments he was at home alone.

This is how I came to understand the way in which metaphors of death and dying, and their equivalents of loss—"withering away", "becoming past-tense", "dying off"—came up in the narratives of my interlocutors living with dementia, and in the stories of many others (elderly and young, with and without dementia) who commented on the accounts presented here. Just like doing nothing could result in metaphorical death, going out and returning home could be a way of fostering the responsive engagement that most of us, for most of our lives, exist in. The accounts of the people with dementia I worked with stood out from the stories of the others whom I spoke to due to the magnitude of the unsettling capacity of the stillness that they found themselves confronted with. If not an outright attempt at "trying not to die", repeatedly going out might have been their way to live.

As others working with Waldenfels' framework have pointed out, the demand placed on us by disturbing forms of alterity can never be entirely tamed (see Louw 2017). This fact both explains the ambiguous and ongoing productivity of unsettling moments (Leistle 2017) and highlights how the support of new and fragile forms of relating opened up in these moments depends on the relational work done by others joining in on a responsive community of care (Dyring and Grøn, n.d.). Following the anxious moment in which she had questioned who indeed she was, Mrs Kelk ran into care worker Celina, who called to her from the other end of the corridor: "Lenie, sweetheart! How are you?" "Do you think I am a sweetheart?" I heard Mrs Kelk mumbling as she slowly walked towards Celina, her left hand following the bar along the wall. Celina softly took Mrs Kelk's arm. A moment later, I found them standing in front of a large picture frame displaying photographs of Mrs Kelk's family members. Celina's fingers were gliding across the images as Mrs Kelk's lips formed the names of her parents, siblings, partner and children. Her tense shoulders relaxed. We left Mrs Kelk sitting on the couch with a warm drink a little later. I hoped that she would be conversant with life for at least a while longer.

Learning About Mobility from Studying the Still

This chapter has explored the ambiguous process whereby experiences of immobility encouraged recurrent practices of leaving and returning home among single-living people with early-stage dementia in a city in the Netherlands. In the accounts presented here, "stillness" (Bissell and Fuller 2011) is a process whereby the everyday self-evidence of seeing, hearing, doing and knowing things is interrupted. Drawing on critical phenomenology (Dyring and Grøn, n.d.) and an anthropology of responsiveness (Waldenfels 2011) has helped me to rethink practices of going out as ways of honing the responsive engagement life exists in.

I have drawn on first-person experiences in order to take issue with approaches that trace mobility practices among people with dementia to disease processes. Instead of directing our attention towards brains in need of dynamic stimuli (Van der Plaats and De Boer 2014), the approach I adopt suggests that we may try to understand how people with dementia who are involved in recursive forms of mobility are bothered by the encompassing senses of social, structural and existential dis-embeddedness that all of us may be familiar with in some way. We may also recognize how such practices are not characterized by the lack of purpose and meaning that is often attributed to "wandering" (Joller et al. 2013), but how they could be quests for support in fostering responsiveness as such.

My rethinking of recursive mobilities as ways of honing existence was enabled by a phenomenological approach which takes "responsiveness [to be] the name for the way to cope with one's fragile and finite existence which captures our being called upon to act anyway" (Wentzer 2018, 215). I proposed that using this approach, which reconceptualizes the human way of being-in-the-world as a process of responsiveness, does more than highlight in new ways how mobility and immobility are intertwined. It may help us to exchange our practices of signifying and objectifying the other for attempts to join the people we work with in dealing with the predicaments of life that many of us find ourselves confronted with (see Leistle 2017, 8–9, 40). In doing so, we may turn our attention towards vulnerability in all of our lives. I would, here, like to take a cue from Bernard Leistle's suggestion that anthropology could be understood as a form of responding (Leistle 2017, 40). If anthropology could be conceived of as a form of responding, perhaps for life and caregiving, we may say the same. In a way not unlike the attentive encounter between Celina and Mrs Kelk, care may then be about honing our abilities to respond.

References

Alzheimer Nederland. 2019. "Factsheet cijfers en feiten over dementie." Accessed December 10, 2019. https://www.alzheimer-nederland.nl/factsheet-cijfers-en-feiten-over-dementie.
Bissell, D. 2008. "Animating Suspension: Waiting for Mobilities." *Mobilities* 2 (2): 277–298.
Bissell, D., and G. Fuller. 2011. *Stillness in a Mobile World*. London: Routledge.
Brittain, K., C. Degnen, G. Gibson, C. Dickinson, and L. Robinson. 2017. "When Walking Becomes Wandering: Representing the Fear of the Fourth Age." *Sociology of Health and Illness* 39 (2): 270–284.
Cipriani, G., C. Lucetti, A. Nuti, and S. Danti. 2014. "Wandering and Dementia." *Psychogeriatrics* 14 (2): 135–142.
Conradson, D. 2011. "The Orchestration of Feeling: Stillness, Spirituality and Places of Retreat." In *Stillness in a Mobile World*, edited by D. Bissell and G. Fuller, 209–228. London: Routledge.
Dewing, J. 2006. "Wandering into the Future: Reconceptualizing Wandering 'a Natural and Good Thing.'" *International Journal of Older People Nursing* 1 (4): 239–249.
Driessen, A. E., I. Van der Klift, and K. Krause. 2018. "Freedom in Dementia Care? On Becoming Better Bound to the Nursing Home." *Etnofoor* 29 (1): 29–41.
Dyring, R. 2015. "A Spectacle of Disappearance." *Tropos* 8 (1): 11–33.
———. 2018. "The Provocation of Freedom." In *Moral Engines: Exploring the Ethical Drives in Human Life*, edited by C. Mattingly, R. Dyring, M. Louw, and T. Schwartz Wentzer, 211–229. Oxford and New York: Berghann.
Dyring, R., and L. Grøn. n.d. "Else and the Little One: A Critical Phenomenology of Potentiality and Life with Dementia.".
Dyring, R., C. Mattingly, and M. Louw. 2018. "The Question of 'Moral Engines': Introducing a Philosophical Anthropological Dialogue." In *Moral Engines: Exploring the Ethical Drives in Human Life*, edited by C. Mattingly, R. Dyring, M. Louw, and T. Schwartz Wentzer, 9–39. Oxford and New York: Berghann.
Ehn, B., and O. Löfgren. 2010. *The Secret World of Doing Nothing*. Berkeley: University of California Press.
Guenther, L. 2013. *Solitary Confinement: Social Death and Its Afterlives*. Minneapolis: University of Minnesota Press.

Hage, G. 2005. "A Not so Multi-sited Ethnography of a Not so Imagined Community." *Anthropological Theory* 5 (4): 463–475.

———. 2009. "Waiting Out the Crisis: On Stuckedness and Governmentality." In *Waiting*, edited by G. Hage, 97–106. Melbourne: Melbourne University Press.

Harrison, P. 2008. "Corporeal Remains: Vulnerability, Proximity, and Living on After the End of the World." *Environment and planning* 40 (2): 423–445.

———. 2011. "The Broken Thread: On Being Still." In *Stillness in a Mobile World*, edited by D. Bissell and G. Fuller, 209–228. London: Routledge.

Higgs, P., and C. Gilleard. 2015. *Rethinking Old Age: Theorising the Fourth Age*. London: Palgrave.

Janeja, M. K., and A. Bandak. (2018). *Ethnographies of Waiting: Doubt, Hope and Uncertainty*. London: Bloomsbury.

Joller, P., N. Gupta, D. P. Seitz, C. Frank, M. Gibson, and S. S. Gill. 2013. "Approach to Inappropriate Sexual Behaviour in People with Dementia." *Canadian Family Physician* 59 (3): 255–260.

Leistle, B. 2014. "From the Alien to the Other: Steps Toward a Phenomenological Theory of Spirit Possession." *Anthropology of Consciousness* 25 (1): 53–90.

———. 2017. "Introduction." In *Anthropology and Alterity: Responding to the Other*, edited by B. Leistle, 1–23. New York and London: Routledge.

Louw, M. 2017. "Burdening Visions: The Haunting of the Unseen in Bishkek, Kyrgyzstan." Contemporary Islam. August 2017.

McDuff, J. and A. Phinney. 2015. "Walking with Meaning: Subjective Experiences of Physical Activity in Dementia." *Global Qualitative Nursing Research* 9 (1): 1–9.

Solomon, O., and M. C. Lawlor. 2018. "Beyond V40.31: Narrative Phenomenology of Wandering in Autism and Dementia." *Culture, Medicine, and Psychiatry*, Published January 24.

Stewart, K. 2007. *Ordinary Affects*. Durham and London: Duke University Press.

Taylor, J. 2017. "Engaging with Dementia: Moral Experiments in Art and Friendship." *Culture, Medicine and Psychiatry* 41 (2): 284–303.

The, B. A. M. 2017. *Dagelijks leven met dementie. Omgaan met de kwetsbaarheid van het bestaan*. Amsterdam: Thoeris.

Van der Plaats, A., and G. De Boer. 2014. *Het demente brein: omgaan met probleemgedrag*. Gytsjerk: Rekladruk.

van Wijngaarden. E., M. Alma, and B. A. M. The. 2019. "'The Eyes of Others' Are What Really Matters: The Experience of Living with Dementia from an Insider Perspective." *PLoS One* 14 (4): e0214724.

Waldenfels, B. 2011. *Phenomenology of the Alien: Basic Concepts*. Evaston, IL: Northwestern University Press.

Wentzer, T. Schwartz. 2018. "Human, the Responding Being: Considerations Towards a Philosophical Anthropology of Responsiveness." In *Moral Engines: Exploring the Ethical Drives in Human Life*, edited by C. Mattingly, R. Dyring, M. Louw, and T. Schwartz Wentzer, 211–229. Oxford and New York: Berghann.

World Health Organisation. 2018. *Towards a Dementia Plan: A WHO Guide*. Geneva: World Health Organisation.

Young, Y., M. Papenkov, and T. Nakashima. 2018. "Who Is Responsible? A Man with Dementia Wanders from Home, Is Hit by a Train, and Dies." *Journal of the American Medical Directors Association* 19 (7): 563–567.

Zigon, J. 2018. *Disappointment: Toward a Critical Hermeneutics of Worldbuilding*. New York: Fordham University Press.

10

Stories of (Im)Mobility: People Affected by Dementia on an Acute Medical Unit

Pippa Collins

Introduction

People living with dementia are becoming an increasingly important part of the acute hospital population taking up a quarter of hospital beds at any one time (Royal College of Psychiatrists 2013; Public Health England 2015). However, due to known under diagnosing of the condition, the number of admissions is in reality higher. In the tertiary hospital within which this study takes place people with a dementia diagnosis are present on every type of ward. These wards cover a wide range of specialities such as clinical haematology, ophthalmology, neurosurgery and occasionally patients in obstetrics.

This patient group have longer hospital stays, are more likely to be discharged to a care home rather than home, and possibly to be more

P. Collins (✉)
University of Southampton, Southampton, UK
e-mail: pc1e13@soton.ac.uk

dependent cognitively and physically at discharge than prior to admission (Alzheimer's Society 2009). Viewing this phenomenon through a mobilities lens can bring understanding as to how and why this occurs.

It will be discussed that whereas mobility defines the hospital it is immobility of the person that shapes the processes of the medical care they receive. The person is rendered a passive, immobile recipient of care provided by a moving, connected professional working in a hospital that relies on the movement of patients through and out of the physical building. The person arrives at the door of the hospital emergency department, their clothing is replaced by a hospital gown; they are placed upon a trolley and become a patient. The sides of the trolley are raised. The patient moves on the trolley to different departments for investigations. The patient is wheeled on their trolley along corridors to the acute medical unit, slid onto a bed and the bed sides are raised; later the bed is wheeled to a ward, and often another and yet another ward. Within wards beds are shifted around as the hospital dictates, the patient on their bed moves again. Professional staff move around the beds delivering care that is task orientated and time dependent (Featherstone et al. 2019). It is not the mobility of the person that is important, but mobility in the world that surrounds them. This very act of moving through the doors of a hospital brings with it a significant change in power and status. Another person crosses the threshold of the hospital and becomes an empowered highly mobile professional. This person is not paid to stand still; movement is a sign of work.

Having a diagnosis of dementia is associated with an increased likelihood of acute hospitalization (Phelan et al. 2012). Once admitted to hospital, it is well established that low mobility levels are common for older people in hospital (Zisberg and Syn-Hershko 2016) with an average of 20 hours each day spent lying down (Brown et al. 2009) and a median daily step count of only 600 (Lim et al. 2018). Low mobility levels are associated with a decline in independence at discharge (Brown et al. 2004; Zisberg et al. 2011), with half of people aged over 85 admitted to an acute hospital declining in independence (Covinsky et al. 2003). Significantly, the effects of low hospital mobility levels last beyond discharge, leaving people with less independence one month later (Zisberg et al. 2011). When older people are active in hospitals it tends to be either with therapy or when undertaking personal care (Lim et al. 2018). Even at times when opportunities are available for a person to move around such as during visiting times, older patients remain inactive as the dominant hospital culture is for visitors to sit around the bedside (Lim et al. 2018). Are visitors simply responding and adapting to the organizational culture of immobility?

The low mobility levels and functional decline that are particularly evident with older inpatients who have dementia (Pedone et al. 2005) are due in part to the fact that nurses find people with dementia challenging to care for (Alzheimer's Society 2009). Areas of particular concern to nurses can include the perceived risk of falling for such patients when walking around or mobilizing, and not having enough time to spend with patients. The longer a person with dementia is in hospital the worse the effect on the symptoms of dementia and their physical health (Alzheimer's Society 2009). This cognitive and physical decline leads to an increase in dependency at discharge. Lafont et al. (2011) use the term "iatrogenic disability" to describe the avoidable dependence in activities of daily living that can be induced during a hospital admission. With relevance to many people in hospital who live with dementia, they identified three patient characteristics that independently predict functional decline: advanced age, cognitive impairment and some level of dependence in activities of daily living prior to admission.

A further explanation of the low levels of mobility that are experienced by older inpatients who have dementia is that within acute hospitals there is a culture of restraint and restriction (Featherstone et al. 2019). This is due to a culture of risk avoidance in acute hospitals whereby concern about people falling over leads to them being confined to their bed or chair (Tadd et al. 2011). People who have dementia often resist the care that is delivered to them by refusing the everyday necessities of a hospital admission such as food, hydration, medication and personal care (Featherstone et al. 2019). This resistance can manifest in different ways such as physical resistance, verbal resistance or refusal to comply with the timetabled rounds of the ward. Physical resistance can include pushing away trays or equipment, turning away from staff or attempting to stand, walk or leave the ward. Attempts to get up from a chair or out of bed are met with restraint from the care staff by either repeated requests to sit down, raised bed rails, sedation, tightly tucked in sheets, low chairs or tables in front of chairs (ibid.).

Rather than being interpreted as a form of communication of unmet needs, or something to be encouraged to avoid iatrogenic disability, moving around for a person with dementia is often labelled as "wandering" or "at risk of absconding". This interpretation of physical activity as "wandering" rather than walking with a purpose leads to a negative understanding of this form of mobility as a purposeless and negative behaviour (Algase et al. 2007; Dewing 2011). When a person with dementia repeatedly attempts to leave, the ward doors are often locked or disguised. As a result, rendering people immobile becomes the default way of ensuring a person's safety.

These practices of restraint and restriction contrast with theories that state mobility is a form of communication that requires sensitive interpretation. For example, agitation, which can be manifested as repeatedly attempting to walk or move around, is often due to distress caused by pain, thirst or overstimulation in a busy ward environment (Elliot and Adams 2011). By recognizing patients' needs and addressing them, the inappropriate use of sedation can be reduced. However, this approach requires staff who are able to interpret this different form of communication and who have the time to do so.

Further immobility is created by the spatial design of hospitals which minimize the need for movement of the patient. A patient is cared for in or beside their bed; their belongings are in a bedside locker; meals are delivered to the bedside; and for people with dementia who need assistance to move and walk they are often given a bed pan or commode beside their beds rather than taken to a toilet (Tadd et al. 2011; Featherstone et al. 2019). Spatial arrangements exist to negate movement (Cresswell 2006) of the patient and minimize time that is spent on caring tasks. Patients with dementia are expected to remain in their place (in or beside their beds) and mobility is portrayed as a threat or dysfunction (ibid.).

Hospital systems permit predictable and relatively risk-free movements (Urry 2007) of patients by rendering the patient immobile and passive, and engendering the professional with the hyper mobility necessary to deliver a range of tasks in a minimum length of time to a static patient. The patient becomes reconfigured as "bits of scattered information distributed across various 'systems'" (ibid.) whereby different professionals perform "body work" (Cohen 2011; Twigg et al. 2011) on different aspects of the patient. The doctor delivers a diagnosis; the nurse delivers the medication; the care assistant delivers personal care; the physical therapists deliver mobility. No one person attends to the entirety of the patient. This treatment of the patient's body as a material object like any other, physical, malleable and ultimately divisible, ignores the fact that bodies are unitary, communicative and mindful (Cohen 2011). The patient waits, immobile and disempowered for people to come to them.

Yet for a person with dementia the ability to move around and to function as independently as possible is crucially reliant upon social connections and environmental conditions (Marshall 1997). They need people who have time to communicate and time to help in a way that is familiar to them. Hospital culture is about efficiency, movement, speed. The fracturing of professional interactions into multiple parts undermines efforts to treat the patient holistically or provide continuity of care (Cohen 2011). This makes it impossible for professionals to build social connections and relationships with the person who has dementia.

All this mobility is noisy which is stressful for people with dementia (Marshall 1997). Acute wards are loud with conversations, footsteps, equipment alarms, buzzers, rattling trolleys and floor cleaning machinery. Cognitive testing is regularly performed by doctors or psychiatrists within hospital settings and yet noise has been shown to detrimentally effect the outcome of cognitive tests in older adults who scored lower in noisy environments (Dupuis et al. 2016). Allowing testing to take place in a noisy, busy ward further disables a person and can lead to them being considered less cognitively able than they actually are. This in turn can lead to restriction and restraint.

Stuck in Motion

The patient with dementia in an acute hospital simultaneously experiences mobility and immobility; they are "stuck in motion" (Castañeda 2018). The patient is stuck in or beside their bed, where a mobile professional attends to their needs as their work pattern allows; yet hospitals rely on "patient flow"—the movement of the patient through the physical space of the hospital, from the emergency department to the wards and back into the community, thus freeing beds for arriving patients. The patient moves but it is not movement of their own choosing.

The consequences of poor patient flow include patients being admitted as "outliers" to wards that are not suited to managing their care, which may mean they have worse clinical outcomes, and inpatients are moved between wards to make room for newcomers (Tadd et al. 2011). This hospital driven movement of the patient can result in poorer outcomes and negative impacts for people with dementia such as increased disorientation and anxiety (ibid.).

The movement of the patient between physical parts of the hospital is not considered as part of the care processes but is considered in terms of hospital operational efficiency; the mobility of one system is inevitably at the expense of the immobility of another (Bissell 2007). At each point in this journey the patient is considered in their place on the ward, sedentary, fixed and immobile. This immobility or "stuckedness" (Hage 2009) has been normalized within hospitals and accepted as an inevitable experience of an inpatient stay. Stuckedness is by definition a situation where a person suffers from both the absence of choices or alternatives to the situation they are currently in and an inability to grab such alternatives even if they are present (Hage 2009).

Within the noisy and hectic ward the voices of people with dementia are figuratively and often literally silent. The imperative of the busy professional is for a patient to answer questions quickly and to tell a clear, concise linear story enabling a diagnosis of their symptoms and the implementation of a treatment plan. And yet, within the first forty-eight hours of an acute hospital admission a person who finds recall on demand difficult, who is in a stressful environment, who may have linguistic and cognitive challenges and may be disorientated to time, place and situation will have had interactions with and been asked questions by at least twenty-four different people who move in and past their bed or trolley space (personal observation). An inability to accurately answer these questions leads to a person being labelled as a "poor historian", "unable to give a history" or simply "confused". Manifestations of frustration at this questioning and noise can be labelled as "aggression". Care processes are based on the assumption that patients will be able to express their wishes, answer questions, acknowledge the needs of other patients, comply with treatment and move through the system as required (Royal College Nursing 2010). Yet this relies on the professional having enough time to stop, listen, understand and, in many cases, observe.

Despite what is known about the consequences of being in hospital for people with dementia, there are very few studies upon which recommendations could be based that have explored the perspectives and experiences of people with dementia during a hospital admission (Gladman et al. 2012; Dewing and Dijk 2016). Qualitative explorations of dementia have tended to involve interviews (Cowdell 2010; Bartlett 2012; Clarke and Bailey 2016; Digby et al. 2017) and whilst these have elicited valuable insights, unilateral questioning can be difficult and threatening for a person who has difficulty recalling events on demand and people can be excluded on the basis of standardized cognitive tests (Digby and Bloomer 2014). Research thus becomes exclusionary when data collection methods rely heavily on intact verbal skills and memory recall which are known to deteriorate with dementia (Bartlett and O'Conner 2010). Research has focused on people who are able to narrate their stories along the classical temporal and contextual lines, and there are therefore few studies that take into account the perspectives and experiences of people living with dementia when they become patients in hospital (Dewing and Djik 2016). Additionally, Bartlett and O'Connor (2010) argue that research that privileges methods that limit or discount the power of the voices of people with dementia need to be challenged.

To help address this issue and to include more people in dementia research, Dewing (2002) argues for the development of dementia specific interview methods. However, I would argue that for research to be inclusive, interview methods should not be used; instead, we should utilize conversations that are led by the person, are in the context of that which is being explored, and we as researchers should learn to listen to the meaning behind the language.

Inclusionary Methods

The research reported here is based upon video recordings of conversations with people with dementia whilst they were inpatients on an acute medical unit. The participants spoke about what was important to them at that time and questions were only asked if they were contextual to the conversation. The videos enabled the capture and critical interrogation of the complex, detailed minutiae of the micromobilities of the participants. Although videos made with and by people who have dementia have been used to empower people to tell their own stories and to enable people who work, live and socialize with them to listen (Capstick 2009, 2011, 2012; Capstick et al. 2016), this research has taken place in long-term care facilities. To our knowledge no study has used video methodology to explore the experiences of people who have dementia whilst they were inpatients in an acute hospital.

Videos are increasingly being used with hand-held devices such as phones, and made publicly available on social networking platforms. People are allowing their private lives to become public. Within hospitals, television film crews are now regularly making documentaries and accessing people during what are often traumatic or emotionally disturbing events. Despite this, videos have not been used for research within acute hospitals with people with dementia. Video can capture a wide range of response possibilities allowing a greater depth of data to be collected. Additionally, the audio-visual materials that are produced can be commented on by participants to produce a further layer of knowledge (Pink 2013) and help create meaningful participation in data creation. In this project, visual methods were used to capture the sense and experience of movement and mobility of the participant, the researcher and of the ward around them, as well as the communication that happens between the participant, the researcher and others.

Visual methodologies can make a major contribution to research due to the egalitarian stance inherent in working alongside the most vulnerable, underrepresented, and least researched and understood members of society (Prosser 2013)—in this case people with dementia in hospital. People with dementia are habitually excluded from research because of the assumption that they are insufficiently articulate to either consent to participate or contribute through interviews or survey samples (Prosser 2013). With particular relevance to the research reported on here, one of the strengths of visual research is that it can record a wide range of possible communication strategies and reveal important information that text or word-based methods cannot (ibid.).

For this research, a tablet with a large screen and good resolution was used to make instant reviewing of the video possible. This was important because it allowed contextual discussion of the research and allowed the participant to comment on their video and decide whether or not to keep it. During the consent process, I showed videos of my family and I having conversations which helped the participants understand what was involved in the research. This contextual conversation about what I was hoping to achieve enabled most participants to consent for themselves. Furthermore, the process of consent was divided into two parts: consent to take part in a videoed conversation and consent to use the videos for research and education. If during the making of the video it became apparent that the participant did not understand that the video was being made for research and education then a family member was consulted. At the end of the video we reviewed the recording together to decide if it should be kept or deleted and to comment upon it. All participants agreed to keep their recordings.

Patient Stories

People with dementia often tell non-contextual stories that are non-linear or entangled within themselves or with the stories of the listener, creating "entangled narratives" (Hydén 2018). Additionally, the person as narrator may move freely through time reliving previous roles such as child, sibling or parent. Their stories become more difficult for the listener to follow and help co-construct due to narrative and historical inconsistencies. These inconsistencies can also lead to the listener not believing in the story and judging the narrator as an unreliable source. Furthermore, for people living with dementia the difficulties encountered with expressive language and loss of memory for recent events and disorientation to place and time may limit the possibility of engaging narratively with the world and with others (Baldwin 2006). Narrative connections are not made, and labels such as "confused" are applied with resulting disempowerment as people cease to listen to the speaker and conversations and decisions are made without them.

Take for example the entangled narrative of Elizabeth. Elizabeth is videoed as she sits in her chair beside her bed. She is beautifully dressed and has the newspaper on her lap. She has been in hospital for two days because of a chest infection which makes her a bit breathless as she talks. All around her is the noise and movement of the busy acute medical unit. In front of her is the nurses' station where there are currently fourteen doctors, nurses and therapists; the conversations are loud. Two people are on the phone. Alarms ring, other patients call out, people pass by her bed as they move between the different parts of the unit. It is noisy and sometimes her voice is nearly drowned out. Her eyes shift constantly to her left to follow the movements of people and her concentration starts to falter as the noise increases to a crescendo.

Elizabeth is talking to the author (P) about her daughters who, since moving back from France, she now sees more regularly. Elizabeth's husband had recently died in France. Elizabeth speaks in long entangled sections, with one thought running into the next and few pauses; the extract below has been divided up to improve intelligibility:

E: OK. They know that I um I don't expect them to be in and out endlessly but they always make a regular call or if they don't come they call me it's so everyday I'm OK

and er because I they know why we moved to France to be different and enjoying so it's doesn't worry them or make them feel they've got to put themselves under my doorstep as it were... because.

P: I understand, I understand. So they like to know that you're close but they don't feel they've to be with you all the time

E: yes no no not have

I mean they're very good they like to take me away going for a walk, going shopping or something like that they'll always pick me up but um

they know that I don't expect or even really want to make them do it I mean at my age you're not buying things all the time are you apart from some bread or something nice to eat for a change.

P: you sound like a very independent resourceful lady
E: well I am yes.
   My my father he he um was missing a long time in the war and my mother she had been a nanny and er we were living and um [names place] and she would always …walk us out to you know but we were never forced to do anything like that but we just …but they were very very caring but not um occupying me all the time you know
   they're not feeling they had to because they wasn't
   no my husband and I we took ourselves to France because that's what we wanted
   so you we could hardly say we we're back now so lets go and had to when how we were cos you can't do that
P: no you can't
E: but er they're very kind and but they don't feel they they've got to pop in at all every day and that's not what I want of them and they've got their other things children and grandchildren and so on.

Elizabeth sits fairly motionless in her chair but her conversation is mobile through time and space as memories appear and are replaced by the next. Elizabeth starts in the here and now, letting the listener know that even though she lives near to her daughters she does not "expect them to be in and out endlessly"; she is an independent, resourceful lady who lived until recently in France with her beloved husband. Elizabeth talks of walks and shopping with her daughters—she might be static now but this is not how she normally is. She travels in time back to her parents and her childhood, how they used to walk out together and then forward in time to France with her husband. Elizabeth is an independent and mobile wife, mother, grandmother and daughter. Yet the extraordinary effort that it takes Elizabeth to excerpt her personhood is not evidenced by her words alone but by close attention to her bodily movements. Elizabeth appears to use her whole body to tell her story. She moves back and forwards in her chair, never relaxing. Her face is mobile and expressive, clearly telling a story as she opens her eyes wide, narrows them or frowns. She uses her hands to indicate words when she appears to have difficulty recalling them or add emphasis to others. It is hard work telling this story amidst the noise and movement of the medical unit. At the end of the conversation, Elizabeth leans back in her chair, her voice hoarse.

Rose is also on the acute medical unit. Her dementia is quite severe which will have affected her cognitive functions such as planning, remembering and language skills. She is in a new and noisy environment and in the last two days has transitioned from her care home to the ambulance, to the emergency department and then to the medical unit where she is now. The background noise is of voices, footsteps, equipment clattering and a phone ringing. People constantly walk past directly in Rose's line of sight. What Rose hears and sees is movement. Sitting beside her bed wrapped in a blanket Rose is not a part of this movement, she sits apart both physically and metaphorically. Rose's movements and words are slow and contrast sharply with the mobility that is all around her—we have our normative language and conventions around body and interactions, but Rose's world is different now.

1 R: [Looks down to right and moves a piece of paper to reveal the roll of tissue]
2 P: Tell me about Manchester
3 R: I can only move very slowly [Keeps head and eyes down as she says this]
5 P: That must be very frustrating
6 R: [swallows twice; is holding the roll of tissue with right hand and focusing on it; moves left shoulder to try to tear off a sheet but arm does not follow. Looks down at the roll and then away into the distance behind me, still holding roll in her right hand but looking away]
7 P: [I can see what she is trying to do with one hand and I take the roll]
8 R: [watches me tearing off tissue]
9 R: I want some paper to blow me nose
10 P: There you go
11 R: [Moves her gaze from downwards to the right, rocks her head twice then looks briefly at me. Blows her nose very carefully using only her right hand, she looks into the middle distance with huge eyes and tiny pupils; slowly puts the tissue down, rests her head on her right hand and rubs her forehead looking down. Raises her head, sniffs and looks directly at me]

12 *P*: Hello?
13 *R*: [Holds my gaze for a moment; puts her head back on her hand and rubs her forehead looking down]
14 *P*: I got you your tea Rose
15 *R*: [Continues rubbing forehead; stops, looks down to the left and right]
16 *R*: Thank you
17 [Stops rubbing her forehead; looks down towards the tea mug with head resting on fingers of hand and to the right then briefly in front and up to right then rests cheek on back of right hand]
18 *R*: Is that tea?
19 *R*: [Looks directly at me briefly then back to the mug and reaches out with her right hand]
20 *P*: It is
21 [Picks up the mug and starts to drink. Looks down at the mug and takes 4 gulps of tea; I mop up a drip and her eyes look down towards her blanket; takes another gulp, she acts as if she is going to put the cup down, I go to take it; she is not and keeps hold of it her eyes watching the mug; takes 4 more gulps whilst looks to the right and back to the mug; I mop another drip; looks down at my hand then at the blanket; glances at me then to right and left quickly]

In order to understand this exchange, Rose's physical body needs to be included. First observations suggest that Rose is not engaging in this conversation, but if the listener/observer slows down to Rose's pace a different impression is given. Rose is fully engaged with this interaction but her conversational partner (P) interrupts her at crucial moments, attempts to micromanage the interaction and outpaces her. P is not giving Rose what she requires the most—time; instead attempting to move the story forwards at P's pace.

Rose has just been given a cup of tea by P and is then requested to "tell me about Manchester" (2). Rose is concerned with the roll of tissue on the Table (1), P is not in her world and attempts to manage the conversation with an out of context comment. What is important to Rose is finding the tissue paper to enable her to blow her nose. She has located the roll of paper and remarks on her slow movements (3) which are further slowed by her left arm not being able to move. Rose continues to focus on the tissue and trying to tear off a sheet. Rose is interrupted in this focused activity with "that must be very frustrating" (5) to which Rose does not respond. The question needs to be asked why P thinks that this must be frustrating to Rose? P appears to have made an automatic assumption that not being able to move quickly, to be apart from the movement and noise around "must be frustrating". Rose does not look frustrated; she is focused on the microquest of blowing her nose. In direct opposition to the surrounding mobility of the ward, Rose is taking her time.

When responded to with the contextual actions of tearing off the piece of tissue and handing it over (7 and 10), Rose says that she wants to blow her nose. For Rose, embodied language appears to be taking a lead, with words in a supporting role. Enactment has assumed a greater role for Rose who has decreased use of words. The verbal narrating of the event has been substituted by bodily enactment (Hydén 2018). This requires the listener to suspend their normal conversational strategies and support Rose in her story. With close attention to Rose's strategies of contextual movements, by slowing the conversation to Rose's pace and using context the conversation can be supported, or scaffolded (Hydén 2018). With contextual actions, and by stopping to listen, Rose can be supported.

Rose becomes entirely focused on blowing her nose and then engages P into her world by looking directly at her (11); to which P responds with "Hello?" This is totally out of context and patronizing Rose; a meaningless intrusion into her world and Rose responds by looking at P and then looking down (13).

Rose is reminded that there is a cup of tea but what is not recognized is that Rose is looking for it on the table which is cluttered and so it is not instantly obvious (17 and 18). It then becomes clear that Rose is extremely thirsty as her movements become quicker as she starts to drink. Verbal conversation is stopped and the two people sit together whilst Rose drinks her tea.

Rose has spoken 19 words in just under three minutes; her embodied movements have said much more. These tell of a microquest to blow her nose, find and drink her tea. However, inattention to her embodied language and not slowing the pace meant that Rose's story was not supported or understood even though Rose is clear what she is doing. To understand Rose, it is necessary to step away from the activity of the ward; to slow down and give notice to the small details. Rose's micromovements are telling a story which is now accessed through the medium of her body; her body has become an active, communicative agent, imbued with its own wisdom, intentionality and purposefulness; her language is embodied (Kontos 2005). Her responses to contextual actions are meaningful because they are bodily manifestations of self that persists despite her severe cognitive impairment (ibid.). However, in order to understand Rose, time is needed to stand still and observe.

When a person is given an opportunity to talk about what is important to them at that time, and when time is made to unconditionally listen, insight can be gained into what it is like being a person who has a diagnosis of dementia. Roger is lying in his bed in hospital pyjamas. Throughout the conversation he is fiddling with the poppers on the pyjama top which don't work. We laugh at this together as it appears that nothing works. Opposite his bed is a critically unwell patient and doctors and nurses move swiftly and frequently past the end of Roger's bed. Roger watches them pass him by as he speaks.

> *P*: Chaotic? I would totally agree with that. I've just been standing and listening………. What I find chaotic was, what I find chaotic is the amount of noise and the stuff going on, what about you?
> *Ro*: Oh yes, that's what it's like, but not only that [coughs], excuse me, its, it's just absolutely chaotic, that's all you can call it. I was due to go home yesterday, now I'm told no you're not going home 'til tomorrow and this is what's happening each day, they keep saying to me no you're not going home today you're going home tomorrow and tomorrow they'll say no you're not going home today you'll be going home tomorrow errr because you can't go home until the doctor has said you are fit to go home, so I've got to be fit to go home but no doctor has come to me so far today to give me any medication which should help me to get out of here. So what do I do? Don't ask me?
> *P*: I, I'm, I feel frustrated just listening to you
> *Ro*: yes well that's the way it is, there's nothing I can do about it, see I mean as far as I know you see cos my memories a bit clouded because I've got Alzheimer's disease but it's not as bad as a lot of them like. As soon as they know you've got Alzheimer's disease a lot of the doctors

"oh well we know all about that" and they don't, they don't know what they're talking about because it's only the person that's got it that knows what it's like. And one of the things about it you see is I am perfectly lucid you know, you listen to me talking.
P: perfectly lucid
Ro: yes yes exactly. Well a doctor will come along he'll read on there umm suffering from Alzheimer's disease and he immediately assumes that you're a nutcase. Do you understand what I mean?

Watching Roger's video is painful. His immobility in bed is a visual manifestation of "stuckedness" (Hage 2009) and embodies his despair at not going home and his powerlessness to affect this. He may have been labelled with a diagnosis of Alzheimer's dementia, but Roger is acutely aware of how this alters other peoples' perceptions of him.

Conclusion

To disregard the stories that people like Elizabeth, Rose and Roger tell is to fail to take into account their side of the hospital mobility/immobility imbalance. A consideration of this imbalance as problematic may serve as a useful way of understanding the experience of being a patient with dementia in a mobile and noisy hospital system. Elizabeth's story moves in time and space; Rose's story is of the micromobilities that make up her quest for a tissue for her nose and a cup of tea. Roger's story is of frustration. What the stories have in common is that each is constructed against a backdrop of other peoples' noise and mobility and each patient represents a different image of "stuckedness". However, time spent attending to the details of each interaction leads to a questioning of the initial impressions of stillness and fixity. Although stuck in one place, each person is using not only verbal language but also embodied language to illustrate their microquests. What is missing is the ability of those around them to attend to these stories. The division of body work between professionals in a striving for hospital efficiency removes attention from the body as a complex, unitary and responsive being, whose needs are labour intensive (Cohen 2011). This division of labour undermines the professionals' abilities to build relationships with patients because they constantly move between patients, and yet it is upon relationships that sensitive and holistic care is built.

The patient at the centre can get lost when the focus of care is on efficiencies and effectiveness, measured only in terms of hospital quality indicators, and which treat the sites of care only insofar as they have an influence on such indicators (Atkinson et al. 2011). New methodologies, such as those reported here, are required to understand issues of movement as they impact on the care of people with dementia in acute hospitals. These methodologies can create an understanding that issues of too little movement or too much, or of the wrong sort at the wrong time (Sheller and Urry 2006) have a direct and significant effect on the person at the centre—the patient. By such understanding it may become possible to address why people with a dementia become less cognitively and physically independent during a hospital admission.

References

Algase, D., D. H. Moore, C. Vandeweerd, and D. Gavin-Dreschnack. 2007. "Mapping the Maze of Terms and Definitions in Dementia-Related Wandering." *Aging & Mental Health* 11 (6): 686–698.

Alzheimer's Society. 2009. *Counting the Cost: Caring for People with Dementia on Hospital Wards*. London: Alzheimer's Society.

Atkinson, S., V. Lawson, and J. Wiles. 2011. "Care of the Body: Spaces of Practice." *Social & Cultural Geography* 12 (6): 563–572. https://doi.org/10.1080/14649365.2011.601238.

Baldwin, C. 2006. "The Narrative Dispossession of People Living with Dementia: Thinking About the Theory and Method of Narrative." In *Narrative, Memory & Knowledge: Representations, Aesthetics, Contexts*, 101–109. Huddersfield: University of Huddersfield.

Bartlett, R. 2012. "Modifying the Diary Interview Method to Research the Lives of People with Dementia." *Qualitative Health Research* 22 (12): 1717–1726.

Bartlett, R., and D. O'Connor. 2010. *Broadening the Dementia Debate: Towards Social Citizenship*. Bristol: The Policy Press.

Bissell, D. 2007. "Animating Suspension: Waiting for Mobilities." *Mobilities* 2 (2): 277–298.
Brown, C., R. J. Friedkin, and S. Inouye. 2004. "Prevalence and Outcomes of Low Mobility in Hospitalized Older Patients." *Journal of the American Geriatrics Society* 52: 1263–1270.
Brown, C., D. Redden, K. Flood, and R. Allman. 2009. "The Under Recognized Epidemic of Low Mobility During Hospitalization of Older Adults." *Journal of the American Geriatric Society* 57: 1660–1665.
Capstick, A. 2009. "This is My Turn; I'm Talking Now: Findings and New Directions from the Ex Memoria Project." *Signpost: Journal of Dementia and Mental Health for Older People* 14 (2): 14–18.
———. 2011. "Travels with a Flipcam: Bringing the Community to People with Dementia in a Day Care Setting Through Visual Technology." *Visual Studies* 26 (2): 142–147.
———. 2012. "Participatory Video and Situated Ethics: Avoiding Disablism." In *The Handbook of Participatory Video*, edited by E.-J. Milne, C. Mitchell, and N. de Lange. Lanham, MD: AltaMira Press.
Capstick A., K. Ludwin, J. Chatwin, and E. Walters E. 2016. "Participatory Video and Well-being in Long-term Care." *Journal of Dementia Care* 24 (1): 26–29.
Castañeda, H. (2018). "'Stuck in Motion': Simultaneous Mobility and Immobility in Migrant Healthcare Along the US-Mexico Border." In *Healthcare in Motion: Immobilities in Health Service Delivery and Access*, edited by C. Vindrola-Padros, G. A. Johnson, and A. E. Pfister, vol. 5, 19–34. New York: Berghahn Books.
Clarke, C., and C. Bailey. 2016. "Narrative Citizenship, Resilience and Inclusion with Dementia: On the Inside or on the Outside of Physical and Social Places." *Dementia* 15 (3): 434–452.
Cohen, R. 2011. "Time, Space and Touch at Work: Body Work and Labour Process (Re)organisation." *Sociology of Health & Illness* 33 (2):189–205.
Covinsky, K., R. Palmer, R. Fortinsky, S. Counsell, A. Stewart, D. Kresevic, C. Burant, and C. Landefeld. 2003. "Loss of Independence in Activities of Daily Living in Older Adults Hospitalized with Medical Illnesses: Increased Vulnerability with Age." *Journal of the American Geriatrics Society* 51: 451–458.
Cowdell, F. 2010. "The Care of Older People with Dementia in Acute Hospitals." *International Journal of Older People Nursing* 5: 83–92.
Cresswell, T. 2006. *On the Move: Mobility in the Modern World*. Routledge: London.

Dewing, J. 2002. "From Ritual to Relationship: A Person-Centred to Consent in Qualitative Research with Older People Who Have Dementia." *Dementia* 1 (2): 157–171.

———. 2011. "Dementia Care: Assess Wander Walking and Apply Strategies." *Nursing and Residential Care* 13 (10): 494–496.

Dewing, J., and S. Dijk. 2016. "What Is the Current State of Care for Older People with Dementia in General Hospitals? A Literature Review." *Dementia* 15 (1): 106–124.

Digby, R., and M. J. Bloomer. 2014. "People with Dementia and the Hospital Environment: The View of Patients and Family Carers." *International Journal of Older People Nursing* 9: 34–43.

Digby, R., S. Lee, and A. Williams. 2017. "The Experience of People with Dementia and Nurses in Hospital: An Integrative Review." *Journal of Clinical Nursing* 26: 1152–1171.

Dupuis, K., V. Marchuk, and M. Pichora-Fuller. 2016. "Noise Affects Performance on the Montreal Cognitive Assessment." *Canadian Journal on Aging/La Revue canadienne du vieillissement* 35 (3): 298–307.

Elliot, R., and J. Adams. 2011. "The Creation of a Dementia Nurse Specialist Role in an Acute General Hospital." *Journal of Psychiatric and Mental Health Nursing* 18: 648–652.

Featherstone, K., A. Northcott, and J. Bridges. 2019. "Routines of Resistance: An Ethnography of the Care of People Living with Dementia in Acute Hospital Wards and Its Consequence." *International Journal of Nursing Studies* 96: 53–60.

Gladman, J., D. Porock, A. Griffiths, P. Clisset, R. Harwood, A. Knight, and F. Jurgens. 2012. "Better Mental Health: Care for Older People with Cognitive Impairment in General Hospitals." Final Report NIHR service Delivery and Organisational Programme.

Hage, G. 2009. "Waiting Out the Crisis: On Stuckedness and Governmentality." In *Waiting*, edited by G. Hage, 97–106. Melbourne: Melbourne University Press.

Hydén, L.-C. 2018. *Entangled Narratives: Collaborative Storytelling and the Re-Imagining of Dementia*. New York: Oxford University Press.

Kontos, P. 2005. "Embodied Selfhood in Alzheimer's Disease: Rethinking Person-Centred Care." *Dementia* 4 (4): 553–570.

LaFont, C., S. Gerard, T. Voisin, M. Pahor, B. Vella and the members of IAGG/AMPA Task force. 2011. "Reducing 'Iatrogenic Disability' in the Hospitalized Frail Elderly." *The Journal of Nutrition, Health & Aging* 15 (8): 645–660.

Lim, S., R. Dodds, D. Bacon, A. Sayer, and H. Roberts. 2018. "Physical Activity Among Hospitalised Older People: Insights from Upper and Lower Limb Accelerometry." *Aging Clinical and Experimental Research* 30 (11): 1363–1369.
Marshall, M. 1997. "Therapeutic Design for People with Dementia." In *Dementia: Challenges and New Directions*, edited by S. Hunter. London: Jessica Kingsley.
Pedone, C., S. Ercolani, M. Catani, D. Maggio, C. Ruggiero. R. Quartesan, U. Senin, P. Mecocci, A. Cherubini on behalf of the GIFA Study Group. 2005. "Elderly Patients with Cognitive Impairment Have a High Risk for Functional Decline During Hospitalization: The GIFA Study." *Journal of Gerontology: Medical Sciences* 60A (12): 1576–1580.
Phelan, E., S. Borson, L. Grothaus, S. Balch, and E. Larson. 2012. "Association of Incident Dementia with Hospitalizations." *Journal of the American Medical Association* 307 (2):165–172.
Pink, S. 2013. *Doing Visual Ethnography*. London: Sage.
Prosser, J. 2013. "Visual Methodology: Towards a More Seeing Research." In *Collecting and Interpreting Qualitative Materials*, edited by N. Denzin and Y. Lincoln, 4th ed. London: Sage.
Public Health England: National Dementia Intelligence Network. 2015. *Reasons Why People with Dementia Are Admitted to Hospital in an Emergency*. PHE publications gateway number: 2014780. http://www.yhpho.org.uk/resource/view.aspx?RID=207311.
Royal College of Nursing. 2010. *Improving Quality of Care for People with Dementia in Hospitals*. London: Royal College of Nursing.
Royal College of Psychiatrists. 2013. *National Audit of Dementia Care in General Hospitals 2012–2013: Second Round Audit Report and Update*. Edited by J. Young, C. Hood, A. Gandesha, and R. Souza. London: HQIP.
Sheller, M., and J. Urry. 2006. "The New Mobilities Paradigm." *Environment and Planning* 38: 207–226.
Tadd, W., A. Hillman, S. Calnan, M. Calnan, T. Bayer, and S. Read. 2011. "Right Place—Wrong Person: Dignity in the Acute Care of Older People." *Quality in Aging and Older Adults* 12 (1): 33–43.
Twigg, J., C. Wolkowitz, R. Cohen, and S. Nettleton. 2011. "Conceptualising Body Work in Health and Social Care." *Sociology of Health & Illness* 33 (2): 171–188.
Urry, J. 2007. *Mobilities*. Cambridge: Polity Press.

Zisberg, A., E. Shadmi, G. Sinoff, N. Gur-Yaish, E. Einav Srulovici, and H. Admi. 2011. "Low Mobility During Hospitalization and Functional Decline in Older Adults." *Journal of the American Geriatric Society* 59: 266–273.

Zisberg, A., and A. Syn-Hershko. 2016. "Factors Related to the Mobility of Hospitalized Older Adults: A Prospective Cohort Study." *Geriatric Nursing* 37: 96–100.

Part III

Motility and (Im)Mobile Possibilities

11

Migratory Labour and the Politics of Prevention: Motility and HPV Vaccination Among Florida Farmworkers

Nolan Kline, Cheryl Vamos, Coralia Vázquez-Otero, Elizabeth Lockhart, Sara K. Proctor, Kristen J. Wells, and Ellen Daley

Introduction

Like many parents might, Blanca, a Latinx (a gender-neutral way to refer to Latino/a) migrant farmworker living in Central Florida, took her sick son to a paediatrician when he felt ill. At the paediatrician's office, Blanca waited for a lengthy period of time before she and her child saw a doctor.

N. Kline (✉) · S. K. Proctor
Rollins College, Winter Park, FL, USA
e-mail: nkline@rollins.edu

C. Vamos · C. Vázquez-Otero · E. Lockhart · E. Daley
University of South Florida College of Public Health,
Tampa, FL, USA

C. Vázquez-Otero
Harvard T.H. Chan School of Public Health,
Dana-Farber Cancer Institute, Boston, MA, USA

K. J. Wells
San Diego State University, San Diego, CA, USA

© The Author(s) 2021
C. Vindrola-Padros et al. (eds.), *Immobility and Medicine*,
https://doi.org/10.1007/978-981-15-4976-2_11

After one hour of waiting and watching the clinic staff chatting, Blanca, increasingly frustrated with the wait, approached a staff member. "They were chatting, and chatting, until I said 'Are you all going to take care of us? We have been here for more than an hour.'" The staff member reprimanded Blanca and told her she arrived without an appointment, so her wait time would be longer. Blanca replied "No, I have an appointment," and the clinic staff member, after verifying Blanca indeed had scheduled the visit, apologized for the error. What the staff member did not realize, however, is that during that time, Blanca potentially risked her livelihood as a migrant farmworker.

Although taking time off from work to go to a doctor can be difficult for many people in the United States (US), for some migrant farmworkers, missing work, even for health-related reasons, can result in job loss. As we describe in this chapter, securing time off of work to take children to the doctor is a significant hardship for many migrant farmworker parents. Although such a routine experience in the US health system—waiting for care—may seem like a minor concern, such moments exacerbate existing health-related vulnerabilities for a highly mobile population that has precarious access to health care and a number of labour-related constraints that limit care for them and their families. Further, labour-related matters can play a role in whether children of migrant farmworkers receive preventive health care, such as receiving the human papillomavirus (HPV) vaccine.

HPV is the most common sexually transmitted infection and causes a number of cancers and genital warts (Viens et al. 2016). Although HPV-related cancers can be prevented through vaccination, there are nevertheless disparities in completion of the two or three shot series. Latinx populations in the US are at increased risk for some HPV-related cancers (Centers for Disease Control and Prevention 2018c), and Latinx adults are less likely to start or complete HPV vaccination (Agénor et al. 2018). Among Latinx immigrants, some subpopulations, such as Latinx migrant farmworkers, experience greater social vulnerabilities related to their economic and political insecurities. Rather than being the consequence of behaviours or individual decisions, these risks are the result of structural factors that require change. In this chapter, we show how agricultural labour expectations and vaccination politics and policies merge to constrain Latinx migrant farmworkers' ability to vaccinate their children for HPV. We argue that labour-related mobility results in a unique form of vaccination motility: never-materializing efforts to complete HPV vaccination for children of migrant farmworkers, despite potential to do so.

Background

Motility and "Stuckness"

As social scientists continue to become attentive to mobility as a theoretical and analytical construct, there has also been growing attention to capacity for mobility. Focusing on the capability for mobility, Hege Høyer Leivestad uses the term "motility" to capture the potential for mobility that may or may not come to fruition (Leivestad 2016). As Leivestad argues, motility offers a way to bridge the gap between mobility and immobility (Leivestad 2016, 134), complicating social and spatial mobility beyond a dichotomous perspective and adding much-needed nuance. Of particular importance to Leivestad is the way in which motility allows for examining temporality and how mobility is "yet-to-be realized, yet-to-be-completed, or never-to-be" (Leivestad 2016, 134). For migrant farmworkers, motility provides a way to examine a population that is significantly spatially mobile because of their labour, while simultaneously experiencing limited social mobility as a result of economic, political, and social subjugation.

Leivestad's notion of motility departs from other social science approaches to the term in that he conceptualizes motility as a concept beyond agentive factors. More than being related to individual choices, for Leivestad, motility is rooted in larger social, cultural, and political phenomena and not limited to notions of individual decisions and behaviour. This emphasis is especially relevant for examining HPV vaccination for migrant farmworkers, who face a number of structural barriers to receiving basic health services. These barriers may result in forms of being "stuck" in a set of circumstances that are structurally determined.

As an analytical device, "stuckness" or the state of being "stuck" in specific social and geographic positions can reveal the types of factors that shape motility. For example, Heide Castañeda shows that some immigrants living along the US and Mexico border feel a sense of being "stuck" or "trapped" because of an increasingly militarized border and the proliferation of checkpoints that aggregate to restrict immigrants' travel for specialized health care. As Castañeda shows, "stuckness" is imposed by larger sociopolitical forces that work together to create health-related exclusions (Castañeda 2018). Further, as Ghassan Hage (2009, 97–107) has argued, "stuckness" can serve as a way to govern populations, resulting in expectations of self-control and "waiting out" the conditions that create "stuckness." The result of being "stuck" in specific social or geographic positions is not by accident, and is instead a reflection of policies that deny sets of rights to certain populations (Auyero 2012).

The Political Economy of Migrant Health

Migrant farmworkers in the United States (US) are a highly mobile population and travel state to state picking and processing produce throughout various growing seasons. National surveys indicate that nearly 83% of migrant farmworkers are Latinx, the majority of whom (69%) were born in Mexico (Hernandez and Gabbard 2018). Overall, farmworkers face a number of health-related vulnerabilities that limit their access to basic preventive care as well as oral and mental health services (Arcury et al. 2015; Saxton 2014, O'Connor et al. 2013; Luque and Castañeda 2013; Kline and Newcomb 2013; Kline 2013; Carrion et al. 2011; Arcury et al. 2009; Arcury and Quandt 2009; Quandt et al. 2007; Lukes and Simon 2006; Early et al. 2006; Arcury et al. 2005; Rao et al. 2003). Many of these vulnerabilities stem from engaging in hazardous work, as farm work is a significant source of injury and exposure to toxic chemicals such as pesticides (Horton 2016; Arcury and Quandt 2009; Quandt et al. 2007). Moreover, many farmworkers have a precarious legal status in the US. Approximately half of farmworkers have work authorization (Hernandez and Gabbard 2018), and if a farmworker is undocumented, they are not eligible for publicly funded health insurance programmes.

Medical anthropologists have demonstrated that many of migrant farmworker' health-related vulnerabilities are rooted in underlying social inequalities. Using a political economy of health perspective that situates health within a system of economic, social, and political circumstances (Singer 1986), we examine how migrant farmworker families' HPV-related health motility hinges on US health policy and migratory labour. Agricultural labour plays a key role in HPV vaccination inequalities, and a political economy focus on migratory labour allows for considering the relationship between work and larger economic and political contexts (Bambra 2011). Part of this political context is the US health insurance system.

In the US, health insurance is often secured through employers, and fewer than one in three migrant farmworkers have employer-provided coverage (Hernandez and Gabbard 2018). Further, in the US's market-based medical system, health services are treated as a commodity for purchase (Rylko-Bauer and Farmer 2002), which is often difficult for migrant farmworkers to obtain since their wages are below national poverty levels. Additionally, migrant farmworkers face a number of obstacles in accessing charitable or subsidized health services due to barriers such as lack of transportation, language challenges, and cost (Arcury and Quandt 2007).

The highly mobile requirements of farm work further constrain migrant workers' ability to receive health care (Wilson et al. 2000; Weathers et al. 2003; Gwyther and Jenkins 1998). Because migrant workers move state to state, they likely encounter multiple health care providers who have inconsistent record keeping and file-sharing practices and may not communicate with one another, limiting follow-up potential (Vamos et al. 2018). Further, rather than simply being a health determinant, there are several determinants to mobility itself, such as aggressive immigration enforcement regimes, that ultimately shape migrants' health-related experiences (Kline 2018a).

While Latinx migrant farmworkers in the US encounter an exclusionary healthcare system and face a number of barriers in accessing care,

US-born children of Latinx migrant farmworkers are eligible for publicly funded health programmes and could theoretically have greater access to services. For example, children of migrant farmworkers can obtain health insurance through the federal childhood health insurance programme and are eligible for Medicaid, the federal health programme for indigent populations. Access to such programmes does not always guarantees services, however, as some research indicates that other constraints, such as providers not accepting Medicaid patients, may further hinder migrant farmworker children's receipt of preventive services (Castañeda et al. 2010). Moreover, in families whose members have different immigration statuses, eligibility for services may not translate to receiving services since undocumented family members may fear their legal status being discovered during a clinical encounter (Castañeda and Melo 2014). In other words, despite potential for care to exist, larger structural forces ultimately restrict types of care that are actually obtainable. These factors are especially salient for considering HPV vaccination for children with migrant farmworker parents.

HPV Vaccination and Migrant Farmworkers

In the US, HPV vaccination rates are lower than other recommended vaccines (Holman et al. 2014). At the time of writing this chapter, the vaccine is recommended for children ages 11–12, can be started at age 9 with catch up shots available through age 26, and licensed for up to age 45 (Centers for Disease Control and Prevention 2018b; Markowitz 2018). Depending on the age of initiation, two or three doses of a vaccine are required to ensure immunity (Centers for Disease Control and Prevention 2018b). Vaccination disparities exist among certain populations, including Latinx groups, and Latinx populations are less likely than non-Latinx whites to begin the HPV vaccine series (Agénor et al. 2018). This is especially concerning given that Latinx populations have elevated risk for some HPV-related cancers that are preventable through vaccination (Centers for Disease Control and Prevention 2018a). Existing literature suggests that possible reasons for differences in vaccination rates among children of Latinx immigrant parents, specifically, include low levels of awareness of HPV and lack of provider recommendation (Aragones et al. 2016). These reasons, however, obfuscate the structural conditions contributing to differences in HPV vaccination rates and individualize the barriers to vaccination that include, as we show in this chapter, farm labour and US vaccination policies. Further, in our research among parents with children who are eligible to receive the HPV vaccine, we found parents had some HPV-related knowledge and awareness, and high levels of vaccine acceptability (Vamos et al. 2018).

The HPV vaccine has a complex legal context in the US since many states do not require children receive the vaccine to enrol in public schools. In part, this is because of the widely debated relationship between HPV and sexual conduct (Colgrove et al. 2010): opposition to mandatory HPV vaccination for school entry has centred around some stakeholders' beliefs that vaccination may lead to promiscuity (Barraza et al. 2016). Lack of mandatory vaccination may partly be responsible for disproportionately low vaccination initiation for vulnerable populations since the HPV vaccine recommendations may not be normalized because they are not part of school entry.

Florida Context

Florida is the third most populous state in the US (United States Census Bureau 2019) and has a robust agricultural economy that relies heavily on migrant agricultural workers. Florida also has a number of adverse sexual and reproductive health outcomes and has one of the nation's top HPV-related cancer rates (Centers for Disease Control and Prevention 2018b). The state's HPV vaccination rates among eligible teens are lower than the national average (Walker et al. 2018), and recent research has shown a significant need to develop vaccination interventions that target the state's largely rural population (Thomas et al. 2019). Combined, the structural barriers and contextual factors shaping HPV vaccination for children of migrant farmworkers demand nuanced attention to examine how such complex factors work together to exacerbate health-related vulnerabilities.

Methods

To understand the structural barriers to HPV vaccination for migrant farmworker families in Central Florida, we conducted three focus groups with migrant farmworker parents (n = 11) who had children aged 9–15 years. We recruited parents in collaboration with a rural faith-based, community organization that serves migrant and seasonal farmworkers. Members of the research team had extensive experience working with the community organization from which we recruited participants, which increased the feasibility of the study since participants likely had precarious immigration statuses and may not have otherwise been willing to participate in the research. We also conducted interviews with key stakeholders, including leaders of organizations who advocate for migrant health issues (n = 6), healthcare providers who see migrant patients (n = 4), a pharmaceutical company representative (n = 1), healthcare administrator (n = 1), and a state legislator (n = 1). We selected study participants by using the social ecological model of health, a framework that situates health in individual, interpersonal, organizational, community-wide, and policy contexts (Stokols 1996), as a theoretical and methodological guide.

The social ecological model of health allowed us to consider how HPV vaccination for children of migrant farmworkers was situated in a larger social context. Accordingly, we used the social ecological model to guide our data collection activities, following other studies on how social factors impact the health of vulnerable Latinx populations (Kline 2019). All interviews and focus groups were conducted in 2015 and professionally transcribed. We used NVivo qualitative data analysis software to thematically code transcripts for barriers and facilitators to HPV vaccination for migrant farmworker families. Overall, the theoretical and methodological approach revealed the structural barriers hindering HPV vaccination for migrant farmworker families, as we described elsewhere (Vamos et al. 2018).

Findings

Missing Work

Findings from parents revealed labour-related challenges with taking children to a doctor's appointment to be vaccinated. For example, in one focus group, parents explained that missing work was not always feasible. "Sometimes at work, they don't give you permission to leave," one parent explained. Another noted that the growing season played a role in an employer's inflexibility. "[It happens more] during strawberry season that they don't let you... you don't leave early; you have to stay until you are done." Another parent added that work schedules can be restrictive, even with designated days off. "Where I work there is an area where they don't let you take time off for appointments or anything. You have to work Monday through Thursday, and [on] Fridays, women don't work Fridays, therefore they have to make their appointments on Fridays because during the other weekdays they cannot ask for permission." During this focus group, one parent noted that such conditions seemed to violate labour laws. "But that is against the law because if your child got sick they would have to give you permission!" Another parent responded by noting that despite potentially being subjected to illegal and clearly questionable labour practices, immigrants' economic needs placed them in such abusive circumstances. The parent explained, "Well, we know [it's illegal], but people who need the job [are the ones who] work there."

Other parents noted that working hours can be inflexible and employers required them to work until the day's work was done. "Sometimes working and living in [employer-provided housing near] the field, sometimes they want you to stay until certain hours; you can't get out just because..." Similarly, one parent noted that work constraints and healthcare providers' hours can be particularly challenging. "The job makes it very difficult because sometimes you can't make it [to doctors' offices] on time or if you make the appointment early in the day and you miss work, the boss doesn't agree, and if you leave work too late [health care providers' offices won't be open]." Another parent added that these conditions can be exacerbated when renting or staying in housing from farm owners: "Sometimes you have to ask for permission to be able to leave because living in a field is harder than living [elsewhere]…Living in a field, if you don't show up for work, they go and knock at the door, and if you are there they take you. If you are not there, they will warn you, and after 2 or 3 times they will fire you." It is not uncommon for migrant farmworkers to live in employer-provided housing, which often has a number of safety and sanitary problems (Arcury et al. 2012), and this type of housing has hidden consequences, such as not being able to stay home with a sick child or take a day off from work to take a child to a preventive care appointment.

While some parents reported labour as a constraint, others found farm work offered them greater flexibility because they could take children to fields with them when they were sick. As one parent noted, "That's why we work in the fields because it's the only place that would let me take care of my children." Further, one parent reported flexibility in taking time off to take children to appointments. "I tell the [boss] 'I have to leave because I have an appointment, I have to leave early…', it's the only job I have found, to be able to work in the fields and to take care of my children." Although some parents noted they did not have as much difficulty leaving work for their children being sick, many parents nevertheless reported how their workplace structured their inability to seek care for children because asking for time off could result in employment termination. Two parents summarized the choice they have to make between their livelihoods and their parental responsibilities, using humour to describe the unfairness of having to choose a job or their children's well-being: "In case of emergency, if the child is dying you cannot let him die!" To which another parent said, "That's why I said, 'if that is the case, you will choose your child, not your job!'"

Mobile Labour and Limited Continuity of Care

Further constraining some parents' ability to get the HPV vaccine for their children are issues related to continuity of care. For example, some parents explained that their Medicaid-eligible children were unable to get Medicaid services in multiple states along the migration route. One parent explained to us how even though her family crossed state lines for work, her children's insurance eligibility did not easily cross the same boundaries. "They, [the clinical staff, will] say 'No that Medicaid doesn't cover here…' They wouldn't take it, [instead], they say 'It doesn't work here because it is from another state, we are not allowed to take it,' so you will have to pay [totally out of pocket] for [a vaccine]." This situation is particularly aggravated by the requirements for HPV vaccines, which necessitate 2–3 doses with the last dose following the first by six months. With a six-month gap, migrant farmworkers could be in a different state by the time the vaccine series is completed from when it was started.

Continuity of care is a problem that one national organization, Migrant Clinicians Network, attempts to address by aiding providers in following and keeping track of migrant patients. One organization leader explained that migrants' highly mobile labour ultimately shapes some providers' approaches to cancer prevention efforts, like screenings:

> I mean, for example, cancer screenings are very interesting because a lot of times a clinician will delay or not even do testing because they know that patient's not going to be there long enough to get the results. And so that's one of the things: making sure the patient is enrolled with us eliminates that reluctance and that reservation… And so you can test them. And if they leave we can be able to provide them with the results and be able to schedule them with any follow-up if necessary.

Accordingly, the network provides a way for clinicians to follow up with patients who may start a screening at one clinic, but who later move due to their labour requirements.

Vaccination Politics and Policies

Not requiring the HPV vaccine for school entry further complicates vaccination for this population. In the US, children's vaccination requirements to attend public schools are created at the state level, and many states, like Florida, do not require HPV vaccination for school entry. One elected official noted that Florida lawmakers likely had no interest in making the vaccine mandatory partly because the majority of state legislators were politically conservative and were less supportive of bills that would add additional regulations or require additional funding for healthcare services. As the lawmaker noted:

> [I'd] say that that probably wouldn't be successful in a Republican dominated house. And the response would be twofold back to a proposal like that. The first response would be, you know, we have problems mandating people do things. And the more nuanced response would be 'but we may come up with a funding source if you want to do an education program to tell them why it's such a good idea'. So I bet you'd have a hard time....Well let me put it to you this way, it would be something we could only do in a nonelection year.

The state legislator further pointed to states with large numbers of conservative voters that had attempted to pass an HPV vaccination mandate, such as Texas, and noted how the electorate negatively responded. Such responses are part of the political context shaping HPV-related vulnerabilities for migrant farmworkers.

Beyond Individual Knowledge

While some existing research on HPV-related prevention among Latinx migrant farmworkers suggests the population may not be knowledgeable about vaccines or willing to get vaccinations, our findings suggested otherwise. During one focus group for example, we asked parents what they knew about the HPV vaccine, and some, though not all, parents readily explained HPV caused cancer in men and women and some parents knew about the vaccine. Moreover, parents expressed acceptance to getting their children vaccinated, or as one parent eloquently explained "why battle [disease] if you can prevent it, right?"

Not only were some parents knowledgeable about vaccines, but health providers also commented on Latinx migrant parents' knowledge and acceptance of vaccines. As one provider explained, "I think our Latino populations are actually the highest vaccine uptakers...I think people have a better understanding of vaccines and a better understanding of the importance of vaccines and also have seen vaccine-preventable diseases…so they understand the dangers of not getting vaccines. We have more work to do with populations who actually are born in the US who don't see these diseases as a threat and don't see these diseases as a possibility…"

Discussion

Overall, findings from our study on the structural determinants to HPV vaccination among migrant farmworker families show how, paradoxically, increased mobility related to agricultural work results in a type of public health "stillness" or "stuckness" in being unable to obtain a preventive public health measure due to larger structural obstacles. This scenario underscores how migrants' HPV-related motility represents potential that has yet to be realized. Migrant children have the potential to be vaccinated because they might have insurance coverage and knowledgeable caregivers willing to vaccinate them. Nevertheless, structural obstacles prevent such efforts from being realized. Such obstacles include labour-related constraints that limit migrant farmworkers' ability take time off work to take children to appointments, demonstrating how HPV prevention efforts are directly tied to economic and labour exploitation.

In addition to labour-related matters, our study further shows how safety net medical programmes in the US, like Medicaid, fail some of the most vulnerable populations. The state-by-state basis on which Medicaid operates limits migrant farmworkers' potential for care continuity, which directly impacts their ability to receive a preventive health measure like the HPV vaccine since it has a temporally important process for dosage. This situation, in particular, reveals how despite being spatially mobile, migrant farmworkers remain in a social space of being "stuck" with ostensible forms of care available but nevertheless out of reach. This type of motility mirrors ways in which publicly funded health programmes that purportedly create safety net services for all populations in the US produce a situation of "false hope" (Castañeda et al. 2010) for children of migrant farmworkers: supposed care that is seemingly accessible but rendered inaccessible through insurmountable structural obstacles.

When put into context with how forms of "stuckness" are rooted in state policies and represent how the state manages rights (Hage 2009; Auyero 2012), migrants' HPV-related motility demonstrates how ostensible access to services is part of a broader sociopolitical effort to assert some migrants' "undeservingness" to certain forms of care (Willen 2011; Yoo 2008). Such matters may worsen in the US as political efforts to constrain some immigrants' rights have specifically been tied to health and social service use. For example, in 2019, the Trump administration proposed broadening existing "public charge" policies that allow for deporting authorized immigrants who rely on governmental assistance to include use of publicly funded services such as Medicaid (Torbati 2019). As existing evidence indicates, these changes may result in some immigrants avoiding certain types of services for themselves or their children, even if they are legally eligible for them (Castañeda and Melo 2014), further solidifying their health-related motility.

The political context surrounding HPV further contributes to migrant farmworkers being unable to get HPV vaccines for children. For example, the lack of an HPV vaccination mandate for school entry at the federal level contributes to the HPV vaccine not being normalized. Such mandates might encounter significant opposition from the US electorate (Abiola et al. 2013; Laugesen et al. 2014), creating little desirability for legislators to propose such laws, as discussed by a policymaker interviewed in this study. These findings demonstrate how the national anxieties about HPV vaccination due to the relationship between HPV and sexual activity have hidden consequences that disproportionately affect marginalized populations. Our study therefore adds to a growing body of research about the hidden health-related consequences of health policies that are not explicitly immigration policies but nevertheless affect immigrants (Kline 2018b).

The situation we have described in this chapter requires attention to address migrant farmworkers' HPV-related vaccination constraints. Solutions to these and similar forms of health inequality directly related to migratory labour, like oral health inequality and cancer disparities, have included a mobility focus, such as bringing mobile dental clinics or mobile screening facilities to fields where migrants work (Luque and Castañeda 2013). Such efforts may be useful for children of migrant farmworkers if mobile clinics go to schools or migrant housing facilities, but these solutions do not address the significant structural issues we have identified here.

In addition to immediate solutions to address migrants' "stuckness" in being unable to get HPV vaccines for children, larger interventions are needed. Such interventions include greater insurance eligibly or changes to Medicaid that reduce existing exclusions and barriers to receiving services, especially when recipients move from state to state. Further, addressing hostile immigration enforcement policies is needed in order to alleviate some immigrants' fears of their immigration status being discovered while obtaining health care, since avoiding detection may lead some parents to avoid obtaining healthcare services for their children. Comprehensive health care for all immigrants regardless of immigration status is therefore necessary.

Health policy solutions must also accompany labour policy solutions. Migrant farmworkers' wage minimums and work conditions are largely unregulated and are exempt from labour laws, a circumstance known as "agricultural exceptionalism" (Horton 2016). As medical anthropologist Sarah Horton (2016) has argued, agricultural exceptionalism must end, and farmworkers, regardless of documentation status, should be granted the same labour rights as other workers. Improving migrant farmworkers' occupational situations could result in improved wages, greater flexibility to take children to preventive care appointments, and an overall reduction in labour exploitation.

Conclusion

In this chapter, we have described a unique form of vaccine-related motility stemming from migratory labour and broader policy contexts. Despite access to health insurance, children of migrant farmworkers may be unable to receive the vaccine because of their parents' highly mobile and insecure labour demands, which exacerbate challenges in taking children to a provider and render some insurance ineffective. Paradoxically, then, this highly mobile population ultimately experiences immobile and inflexible systems of care and labour that complicate preventive health measures. In the aggregate, these factors demonstrate how attention to theorizing mobility rather than understanding mobility simply as spatial movement can provide much-needed nuance on how social factors shape poor health.

References

Abiola, Sara E., James Colgrove, and Michelle M. Mello. 2013. "The Politics of HPV Vaccination Policy Formation in the United States." *Journal of Health Politics, Policy and Law* 38 (4): 645–681.

Agénor, Madina, Ashley E. Pérez, Sarah M. Peitzmeier, and Sonya Borrero. 2018. "Racial/Ethnic Disparities in Human Papillomavirus Vaccination Initiation and Completion among US Women in the Post-Affordable Care Act Era." *Ethnicity & health* 25: 1–15.

Aragones, Abraham, Margaux Genoff, Cynthia Gonzalez, Elyse Shuk, and Francesca Gany. 2016. "HPV Vaccine and Latino Immigrant Parents: If They Offer It, We Will Get It." *Journal of Immigrant and Minority Health* 18 (5): 1060–1065.

Arcury, Thomas A., and Sara A. Quandt. 2007. "Delivery of Health Services to Migrant and Seasonal Farmworkers." *Annual Review of Public Health* 28 (1): 345–363.

———. 2009. *Latino Farmworkers in the Eastern United States: Health, Safety and Justice*. New York: Springer Verlag.

Arcury, Thomas A., Maria Weir, Haiying Chen, Phillip Summers, Lori E. Pelletier, Leonardo Galván, Werner E. Bischoff, Maria C. Mirabelli, and Sara A. Quandt. 2012. "Migrant Farmworker Housing Regulation Violations in North Carolina." *American Journal of Industrial Medicine* 55 (3): 191–204.

Arcury, Thomas A., Sara A. Quandt, Pamela Rao, Alicia M. Doran, Beverly M. Snively, Dana B. Barr, Jane A. Hoppin, and Stephen W. Davis. 2005. "Organophosphate Pesticide Exposure in Farmworker Family Members in Western North Carolina and Virginia: Case Comparisons." *Human Organization* 64 (1): 40–51.

Arcury, Thomas A., Phillip Summers, Jennifer W. Talton, Haiying Chen, Joanne C. Sandberg, Chaya R. Spears Johnson, and Sara A. Quandt.

2015. "Heat Illness Among North Carolina Latino Farmworkers." *Journal of Occupational and Environmental Medicine* 57 (12): 1299–1304.

Arcury, Thomas A., M. F. Wiggins, and Sara A. Quandt. 2009. "An Agenda for Farmworker Social Justice in the Eastern United States." In *Latino Farmworkers in the Eastern United States: Health, Safety and Justice*, edited by T. A. Arcury and S. A. Quandt. New York: Springer.

Auyero, Javier. 2012. *Patients of the State: The Politics of Waiting in Argentina*. Durham and London: Duke University Press.

Bambra, Clare. 2011. "Work, Worklessness and the Political Economy of Health Inequalities." *Journal of Epidemiology & Community Health* 65 (9): 746–750.

Barraza, Leila, Kim Weidenaar, Doug Campos-Outcalt, and Y. Tony Yang. 2016. "Human Papillomavirus and Mandatory Immunization Laws: What Can We Learn from Early Mandates?" *Public Health Reports* 131 (5): 728–731.

Carrion, Iraida V., Heide Castañeda, Dinorah Martinez–Tyson, and Nolan Kline. 2011. "Barriers Impeding Access to Primary Oral Health Care Among Farmworker Families in Central Florida." *Social Work in Health Care* 50 (10): 828–844.

Castañeda, Heide. 2018. "Stuck in Motion": Simultaneous Mobility and Immobility in Migrant Helathcare Along the US-Mexico Border." In *Healthcare in Motion: Immobilities in Health Service Delivery and Access*, edited by Cecilia Vindrola-Padros, Ginger A. Johnson, and Anne E. Pfister, 19–34. New York and Oxford: Berghahn Books.

Castañeda, Heide, Iraida V. Carrion, Nolan Kline, and Dinorah Tyson-Martinez. 2010. "False Hope: Effects of Social Class and Health Policy on Oral Health Inequalities for Migrant Farmworker Families." *Social Science & Medicine* 71 (11): 2028–2037.

Castañeda, Heide, and Milena Andrea Melo. 2014. "Health Care Access for Latino Mixed-Status Families Barriers, Strategies, and Implications for Reform." *American Behavioral Scientist* 58 (14): 1891–1909.

Centers for Disease Control and Prevention. 2018a. "Early Policy Responses to the Human Papillomavirus Vaccine in the United States, 2006–2010." *Journal of Adolescent Health*.

———. 2018b. "HPV-Associated Cancer Rates by State." Accessed April 12. https://www.cdc.gov/cancer/hpv/statistics/state/.

———. 2018c. "HPV-Associated Cancers Rates by Race and Ethnicity." Accessed April 12. https://www.cdc.gov/cancer/hpv/statistics/race.htm.

Colgrove, James, Sara Abiola, and Michelle M. Mello. 2010. "HPV Vaccination Mandates—Lawmaking Amid Political and Scientific Controversy." *New England Journal of Medicine* 363 (8): 785–791.

Early, Julie, Stephen W. Davis, Sara A. Quandt, Pamela Rao, Beverly M. Snively, and Thomas A. Arcury. 2006. "Housing Characteristics of Farmworker Families in North Carolina." *Journal of Immigrant and Minority Health* 8 (2): 173–184.

Gwyther, Marni E., and Melinda Jenkins. 1998. "Migrant Farmworker Children: Health Status, Barriers to Care, and Nursing Innovations in Health Care Delivery." *Journal of Pediatric Health Care* 12 (2): 60–66.

Hage, Ghassan. 2009. *Waiting*. Melbourne: Melbourne University Publishing.

Hernandez, Trish, and Susan Gabbard. 2018. "Findings from the National Agricultural Workers Survey (NAWS) 2015–2016: A Demographic and Employment Profile of United States Farmworkers." U.S. Department of Labor, Employment and Training Administration.

Holman, Dawn M., Vicki Benard, Katherine B. Roland, Meg Watson, Nicole Liddon, and Shannon Stokley. 2014. "Barriers to Human Papillomavirus Vaccination Among US Adolescents: A Systematic Review of the Literature." *JAMA Pediatrics* 168 (1): 76–82.

Horton, Sarah Bronwen. 2016. *They Leave Their Kidneys in the Fields: Illness, Injury, and Illegality Among US Farmworkers*. Vol. 40. Oakland, CA: University of California Press.

Kline, Nolan. 2013. "'There's Nowhere I Can Go to Get Help, and I Have Tooth Pain Right Now': The Oral Health Syndemic Among Migrant Farmworkers in Florida." *Annals of Anthropological Practice* 36 (2): 387–401.

———. 2018a. "'It's Too Risky to Leave the House': Immigrant Policing and Health-Related Mobility." In *Healthcare in Motion: Immobilities in Health Service Delivery and Access*, edited by Cecilia Vindrola-Padros, Ginger A. Johnson, and Anne E. Pfister, 35–52. New York and Oxford: Berghahn Books.

———. 2018b. "Life, Death, and Dialysis: Medical Repatriation and Liminal Life Among Undocumented Kidney Failure Patients in the United States." *PoLAR: Political and Legal Anthropology Review* 41 (2): 216–230.

———. 2019. *Pathogenic Policing: Immigration Enforcement and Health in the US South*. New Brunswick, NJ: Rutgers University Press.

Kline, Nolan, and Rachel Newcomb. 2013. "The Forgotten Farmworkers of Apopka, Florida: Prospects for Collaborative Research and Activism to Assist African American Former Farmworkers." *Anthropology and Humanism* 38 (2): 160–176.

Laugesen, Miriam J., Ritesh Mistry, Kelley A. Carameli, Kurt M. Ribisl, Jack Needleman, and Roshan Bastani. 2014. "Early Policy Responses to the Human Papillomavirus Vaccine in the United States, 2006–2010." *Journal of Adolescent Health* 55 (5): 659–664.

Leivestad, Hege Høyer. 2016. "Motility." In *Keywords of Mobility: Critical engAgements*, edited by Noel B. Salazar and Kiran Jayaram, 133–151. New York: Berghahn Books.

Lukes, Sherri M., and Bret Simon. 2006. "Dental Services for Migrant and Seasonal Farmworkers in US Community/Migrant Health Centers." *The Journal of Rural Health* 22 (3): 269–272.

Luque, John S., and Heide Castañeda. 2013. "Delivery of Mobile Clinic Services to Migrant snd Seasonal Farmworkers: A Review of Practice Models for Community-Academic Partnerships." *Journal of Community Health* 38 (2): 397–407.

Markowitz, Lauri E. 2018. "Expanded Age Range for 9-Valent HPV Vaccine: Background for Policy Considerations."

O'Connor, Kathleen, Maria Stoecklin-Marois, and Marc B. Schenker. 2013. "Examining Nervios Among Immigrant Male Farmworkers in the MICASA Study: Sociodemographics, Housing Conditions and Psychosocial Factors." *Journal of Immigrant and Minority Health* 17 (1): 198–207.

Quandt, Sara A., Heather M. Clark, Pamela Rao, and Thomas A. Arcury. 2007. "Oral Health of Children and Adults in Latino Migrant and Seasonal Farmworker Families." *Journal of Immigrant Minority Health* 9: 229–235.

Rao, P., T. A. Arcury, and S. A. Quandt. 2003. "A Culturally Appropriate Farmworker Pesticide Safety Program." *Practicing anthropology* 25 (1): 10–13.

Rylko-Bauer, Barbara, and Paul Farmer. 2002. "Managed Care or Managed Inequality? A Call for Critiques of Market-Based Medicine." *Medical Anthropology Quarterly* 16 (4): 476–502.

Saxton, Dvera I. 2014. "Strawberry Fields as Extreme Environments: The Ecobiopolitics of Farmworker Health." *Medical Anthropology* 34 (2): 166–183. https://doi.org/10.1080/01459740.2014.959167.

Singer, Merrill. 1986. "Developing a Critical Perspective in Medical Anthropology." *Anthropology Quarterly* 17 (5): 128–129.

Stokols, Daniel. 1996. "Translating Social Ecological Theory into Guidelines for Community Health Promotion." *American Journal of Health Promotion* 10 (4): 282–298.

Thomas, Tami L., Michelle Caldera, and Jeffrey Maurer. 2019. "A Short Report: Parents HPV Vaccine Knowledge in Rural South Florida." *Human Vaccines & Immunotherapeutics* 15: 1–6.

Torbati, Yeganeh. 2019. Exclusive: Trump Administration Proposal Would Make It Easier to Deport Immigrants Who Use Public Benefits. *Reuters*. Accessed July 11, 2019.

United States Census Bureau. 2019. "Annual Estimates of the Resident Population: April 1, 2010 to July 1, 2018." Accessed June 25. https://factfinder.census.gov/faces/tableservices/jsf/pages/productview.xhtml?src=bkmk.

Vamos, Cheryl A., Coralia Vázquez-Otero, Nolan Kline, Elizabeth A. Lockhart, Kristen J. Wells, Sara Proctor, Cathy D. Meade, and Ellen M. Daley. 2018. "Multi-Level Determinants to HPV Vaccination Among Hispanic Farmworker Families in Florida." *Ethnicity & health*: 1–18.

Viens, L. J., S. J. Henley, M. Watson, L. E. Markowitz, C. C. Thomas, and T. D. Thompson. 2016. "Human Papillomavirus-Associated Cancers—United States, 2008–2012." *Morbidity and Mortality Weekly Report* 65 (26): 661–666.

Walker, Tanja Y., Laurie D. Elam-Evans, David Yankey, Lauri E. Markowitz, Charnetta L. Williams, Sarah A. Mbaeyi, Benjamin Fredua, and Shannon Stokley. 2018. "National, Regional, State, and Selected Local Area Vaccination Coverage among Adolescents Aged 13–17 years—United States, 2017." *Morbidity and Mortality Weekly Report* 67 (33): 909.

Weathers, Andrea, Cynthia Minkovitz, Patricia O'Campo, and Marie Diener-West. 2003. "Health Services Use by Children of Migratory Agricultural Workers: Exploring the Role of Need for Care." *Pediatrics* 111 (5): 956–963.

Willen, Sarah S. 2011. "How Is Health-Related 'Deservingness' Reckoned? Perspectives from Unauthorized Im/Migrants in Tel Aviv." *Social Science & Medicine* 74 (6): 812–821.

Wilson, Astrid Hellier, Judith Lupo Wold, Lorine Spencer, and Kathleen Pittman. 2000. "Primary Health Care for Hispanic Children of Migrant Farm Workers." *Journal of Pediatric Health Care* 14 (5): 209–215.

Yoo, Grace J. 2008. "Immigrants and Welfare: Policy Constructions of Deservingness." *Journal of Immigrant & Refugee Studies* 6 (4): 490–507.

12

Living Suspended: Anticipation and Resistance in Brain Cancer

Henry Llewellyn and Paul Higgs

Introduction

Cancerous brain tumours are by medical definition 'progressive.' They change, often rapidly, disrupt, invade and infiltrate people's bodies imposing continued physical and mental debility. So too do the social formations of care change—institutional and familial. Neuro-oncology—the clinical subfield which designates and manages cancer of the brain—is currently in a period of major disruption as new molecular genetic biomarkers revise diagnostic categories, mark new entries and closures to clinical trials, and increasingly determine treatment decisions in standard care settings (Louis et al. 2016). Patients and those around them find new ways of relating, forming new patient-to-family, patient-to-clinician and patient-to-patient alliances which engender new forms of dependence and possibility. These various currents of change—bodily, technical and relational—intersect and interact in often unpredictable

H. Llewellyn (✉) · P. Higgs
University College London, London, UK
e-mail: h.llewellyn@ucl.ac.uk

ways, ushering forth new realities and potential futures with sudden and brutal force. Patients, families and clinicians navigate terrains in continual formation (Llewellyn et al. 2018).

However, moments of stasis or relative calm exist amid this flux—when transitions pass or progressions slow, when knowledge platforms approach standard care and relationships solidify. This chapter focuses on these moments. In particular, it focuses on patients' subjectivities and conceptions of future when living in suspension. It draws attention to the nervous anticipations that often mark these waits; it focuses on the assembling of new imagined possibilities of what is to come and the false dawns of change. Indeed, given the peculiar condition of living with the prognosis of a disease which affects the brain, patients often mistake the imagined for the real. For people with a brain tumour, these imagined possibilities entail images of becoming lost to themselves and others through what is clinically and legally described as mental incapacity.

Drawing on ethnographic field research (2014–2016) in the UK among people diagnosed with aggressive—fast growing or 'high grade'—brain tumours, we ask: What does it mean to live suspended? How do waiting and suspension relate to the potential and the possible? What forms of possibility emerge and how do these connect to the rubrics of institutions that seek to name, diagnose and 'fix'? What are the embodied and affective dimensions of waiting when the awaited is a terrible mystery, impossible to locate or predict?

Ethnographically, we chart several features of patients' experiences and the social formations through which these experiences unfold: first, the images of stability given in technologies such as radiographical images, clinical consultations and the management categories of 'watch and wait'; second, the common images of mind, brain, and injury which prime those afflicted to doubt; and third, the imagined losses of self that characterize the lives of people with a progressive brain tumour even though they might be rendered 'stable' by medicine. In this latter project, we document patients' encounters with strangeness; how these encounters are interpreted as harbingers of a new reality; and how patients and families seek to delay this fate.

Our theoretical treatment is informed by recent ideas about possibility and emergence articulated in the anthropologies of potentiality (Taussig et al. 2013) and becoming (Biehl and Locke 2010, 2017). Though differing in emphasis and application, collectively, these call for analyses attuned to the patterns of emergence and the affective states they inspire: they direct us to consider how current events, anticipated futures and subjectivities are framed and entangled. They underscore the always provisional nature of reality and sense of imminence which accompanies moments poised for radical transition. Important to the current project is how such approaches draw distinctions and relations between the projected and the lived, and the recursive nature through which the future is informed by the present and the present informed by the future. They also invite consideration of the different registers that locate things as stable or in motion, and indeed the intersubjective experience of the unknown and the anticipated (Stephan and Flaherty 2019).

We find further—and related—theoretical inspiration in the emerging scholarship on waiting (Hage 2009; Janeja and Bandak 2018; Mattingly 2019). This has included efforts to consolidate an anthropology of waiting and forward a conceptual framework of its political, relational and existential dimensions (e.g. Ehn and Löfgren 2010; Janeja and Bandak 2018). We find recent attention on the structuring of waiting experience especially useful along with analyses attuned to what have been described as agentive or *active* modes of waiting (e.g. Appadurai 2013; Dwyer 2009). These have highlighted both the institutional conditions which force people to wait and the dynamic and affective orientations towards realization or, in Appadurai's (2013) formulation, a 'politics of patience.' Ethnographically, scholars have thus documented the structural violence inherent in a politics of waiting, wherein socially marginalized groups—e.g. refugees (Andersson 2014), prisoners (Reed 2011) or the poor (Auyero 2012)—are compelled to wait, as well as the tactical ways they might circumvent delay to enact their imagined trajectories (e.g. Petryna 2002). In this article, we contribute to this notion of active waiting by foregrounding neither hopeful forms of waiting nor those modes through which people seek to realize imagined futures more quickly, but the tactics of delay or postponement improvised by patients; for the imagined futures we highlight in this article are deeply terrifying, profoundly unwanted and utterly unknowable. After a brief description of fieldwork, we discuss the institutional forms and popular images of brain tumours which mark time and produce an 'imaginative space of potentiality' (Taussig et al. 2013) for those affected. We then describe how these images become entwined in patients' conceptions of themselves and further how they are resisted. We end with a discussion of what we call an 'anticipatory loss of self' as a hallmark feature of brain tumour experience.

Fieldwork

The fieldwork informing this chapter was conducted by Llewellyn over 18-months in a specialist hospital for people with neurological disease in the UK. More broadly, this fieldwork focused on the production and application of biomedical knowledge about tumour diagnosis and treatment and the complex processes of medical decision-making. It involved long-term participant observation in clinical and domestic settings—including clinics, laboratories, radiology departments, surgical theatres, lecture halls, waiting rooms and patients' homes—and repeated semi-structured and unstructured interviews with patients, families and clinicians. It also included analysis of key scientific articles and policy on standard care practice and innovation. We have kept the first-person account in our presentation of ethnographic data to convey the relational qualities of fieldwork with greater fidelity and intensity. Names given are pseudonyms.

Uncertain Time

Modern medicine seeks to 'fix.' By this, we mean to emphasize the ways in which modern medicine attempts to fix uncertainty and instability in relatively static landscapes as well as the implication to intervene in—or *fix*—the course of disease (Fox 2000; Llewellyn et al. 2018). It is therefore to establish fixity in diagnosis and to arrest or even reverse disease progression, the successes of which are described by clinicians as making patients 'stable.'

By establishing a predictive framework in which to situate the unfolding nature of disease, medicine also fixes time (DelVecchio Good et al. 1994). As such, however imprecisely, it builds a frame around disease and time, establishing a logic of possibility and logic of waiting. Waiting is indeed a state intrinsic to prognosis and it is animated by the specific predictions of debility and decline. It is this which constitutes what Lochlann Jain calls the strange condition of 'living in prognosis' (Jain 2013, 103)—a condition in which futures are severed from the ordinary markers of life-course and reformed within disease's hourglass.

Though intervention is embedded within cancer cultures as near moral imperative (DelVecchio Good et al. 1990), there are periods of non-intervention. The so-called watch and wait draws images of a disease under siege by clinicians prepared to act at any sign of change revealed in symptoms, scans and sometimes the analysis of new tissue samples. It evokes both feelings of imminence and the safety of knowing there are ways of re-establishing stability should change be seen. At times, these periods of waiting follow periods of action, written into the logic of the treatment pathway. They also happen when clinicians are unsure of patients' pathology or how it will mature; whether scans show change or not, helping to build a more detailed diagnosis. Watch and wait is also therefore a means of gathering information.

Yet, despite its sophistication, medicine is unable to give precise time frames. Prognosis is an especially difficult medical practice, part science and part art, and, like all aspects of care, socially contingent. Llewellyn witnessed, for example, how disease progression, the assumed trajectory of disease, is disclosed haphazardly in a progressional ordering of new moments, missed opportunities, and the positive characterizations of growth, spread and transformation. These are laden affectively with the hopes, determinations, doubts, sadness and resignations of patients, families and clinicians. Everything is always provisional and the images of future are images of precarious *potential*.

Primed for Doubt

A brain tumour diagnosis is a dominant frame of understanding for those afflicted; it primes them to doubt. During diagnosis and throughout their care, patients are socialized into seeing themselves anew through repeated scans, which mark out their brains in terms of 'eloquence' and 'essential functioning.' They are sometimes awake during operations and asked to respond to structured questions as surgeons stimulate parts of their brains to localize function and establish what is safe to remove. Some patients watch videos of neurosurgery on YouTube and see with dramatic effect how a touch from a charged probe to the brain makes someone slur their words, stutter or stop shaking their arms. They learn the idiosyncrasies of their brains: the specific neural patterns which shape whether they are left or right-handed, or the precise location of their motor strip, and they are told that this is the site that makes them smile, frown or move their fingers.

Stories in the media do much to dramatize public narratives of the disease and animate patients' concerns, depicting people with brain tumours as pathological, with unnerving symptoms. In a recent profile entitled 'The Neuroscientist who lost her mind,' *The Independent* told the story of neuroscientist Barbara Lipska, who was diagnosed with a brain tumour in 2015. It quotes her saying: 'The tumours in my brain became inflamed, and it was this subsequent swelling that made me lose my mind. I was crazy for probably two months' (2 April 2018). The case of a man with a brain tumour arrested and sentenced to four years in prison for threatening women with a baseball bat made the press under headlines such as 'Brain tumour survivor "wanted to kill blondes",' (*Daily Telegraph*, 27 February 2018) and 'Man with no previous convictions who became obsessed with wanting to kill blonde women following BRAIN TUMOUR surgery is jailed for four years' (MailOnline 26 February 2018, *capitals in original*). In these accounts, it is the image of a mass which appears full force—a mass which invades the brain and also the mind, assuming control of a persons' thoughts, feelings and actions, often with malign intent and towards menacing ends.

Such images further connect patients to the location of self given in a Cartesian materialism, which might be said to prevail in the West (Vidal

2009). The tight binding of mind and brain given in this ontology is, we argue, what grounds patients' conceptions that damage to the brain equates with loss of self. It is an ontology embedded in contemporary neuroscience that also circulates freely in popular culture (Rose and Abi-Rached 2013; Williams et al. 2012). As patients become diagnosed, see images of their brains, watch videos of neurosurgery, become attuned to headlines of sensational losses of mind and so on, these popular ways of locating the self become further reified: the brain further becomes a site of key significance. It is through these imaginative spaces of potentiality that patients are so often primed to doubt current stabilities and assemble their futures of mental incapacity as inevitable ends. To borrow a phrase from The Brain Tumour Charity's report on brain tumour experience (The Brain Tumour Charity 2015), 'losing myself' is a primary fear for patients and for their families, represented as a terrible mystery and an ever-present concern.

Encountering the Strange

Early in fieldwork, I (Llewellyn) met with patients turned blind by tumours that had crushed their optic nerves, who had major problems speaking, who remembered little beyond the simplest details of things, who forgot the names of their families or who exhibited the excesses of hormones like growth hormone because their pituitary glands had been radically disturbed. Not all these afflictions affected the patients with whom I made deeper engagements, though all had suffered what they considered strange sensations and experiences like seizures and premonitory auras.[1]

Sara, for example, 72 years old when I met her, began suffering more serious seizures six months after she was diagnosed. Following a particularly heavy seizure late at night, she was rushed to hospital by ambulance

---

[1] Seizures are sudden bursts of electrical activity in the brain which can transpire as jerky movements or perceptual disturbances. They are common in people diagnosed with a brain tumour and can be accompanied by strange or intense feelings called 'auras.' Patients are often prescribed medication to control seizures, though doing so pharmacologically can be extremely challenging.

and spent a week under close observation. She told me she was delirious for most of this seizure and for weeks after being discharged could not place the events in a clear narrative. For her, it was as though time had been lost.

During these weeks, she and Robbie, her husband, came back to the hospital each week for blood tests and to see the oncologist. She had finished radiotherapy months back and was now midway through chemotherapy. We met in the hospital foyer, as usual before her appointment with the oncologist. Her hearing had worsened suddenly after the seizure and she now found herself easily confusing things. On this occasion, the clinic was on the lower ground floor, unlike usual, and this was especially hard for her to reconcile. 'In my head, we should be going up there,' she kept saying, pointing up towards the 4th floor. Consistent with the experiences of almost every patient I met with, these small changes to routine could at times be unsettling in the extreme.

Sara moved slowly through her confusion and the emotional toll it took on her. Strange episodes replayed as she kept thinking she was in her parents' house: 'It's really weird: when I'm upstairs I think I'm with my mum and dad and when I'm downstairs I'm back home with Robbie.' These thoughts would be extremely affective and laden with the details of old memories of her childhood, but her parents had both died more than twelve years ago. She felt out of place and out of time.

I would sometimes ask people to try to tell me what they saw and how they felt—what an aura is like, or a seizure, or indeed what they saw in ordinary moments. Describing her eyesight, which had seriously deteriorated after her surgery, Fay, who was 49 years old when we first met, told me: 'It's so confusing—so disorientating.' This was hugely distressing for her. She lived alone and so these kinds of incapacity were especially significant for her. At first, her oncologist hoped that things would get better with her eyes. He attributed to it to swelling caused by surgery, but instead it continued to deteriorate and Fay was eventually registered partially sighted. When I asked her more specific questions about what she saw, she told me:

> *Fay*: If I'm looking at your face—I can see the whole of your face—but I can't see what is at the side of you, I can't see the door, I can't see the

door opening, I can't see the end of the sofa. I can see the picture—well I know the picture is above your head, I can see the outline of the bottom part—but I can't see if—I can't see the whole of the picture—
Henry: And what about to your right?
Fay: Yeah I can see the television, the TV's there. I can see Dennis's [Fay's cat] blankets on the floor. I can see the chair.
Henry: And what does it actually look like, I mean the stuff that you can't see, what is there instead?
Fay: Nothing, it's black—it's empty space, it's like it's not there.

Describing auras—perceptual disturbances that may or may not precede a seizure—Jim told me:

It feels just a bit like too much information in your head that you can't process ... this is what would happen to me at work and another thing that I noticed, for over a year before maybe, is that I'd be working on something late into the night and I wouldn't be able to get it out of my head. And that's exactly how the seizure started. It just felt like an overload of information in my head. Like a buzzing sensation and then I just felt the buzzing sensation just take over and then spread from one side of my head to the other. So now what I fear is just that overload of information.

Jim was 35 years old when we met and married to Tina. After the rare event of a stroke during surgery, he had almost completely lost the power down his left side and was unable to walk. He told me how his frequent seizures were difficult to control and would make him feel as though he was losing track of hours and sometimes days. He sometimes fitted at night making it harder for him to place himself in time. Like others, Jim told me about the peculiar feeling of time lost to seizures and, when I first met him shortly after surgery, he relied on Tina to fill in the gaps. He found it hard being in groups, finding conversation difficult to track with, as he put it, 'trouble finding the right words and responding quickly enough: I'm generally just slower at talking and thinking.' Most of all, he was concerned with not making sense and doubting his reality, as this exchange between Jim and Tina, a year later, shows:

Jim: People keep asking me how I feel or do I feel different. I don't know. Or am I doing that? [to Tina directly] You keep questioning my sanity—
Tina: [laughing] I don't— [to me directly] it's because he was having a conversation in his sleep last night, saying that had I paid the tennis instructor—I said, 'You haven't been to tennis.' And he said, 'yes I had.' And I said, 'Maybe it's for Tom'—our [son]—and I said, 'but Jim, you can't walk.' And he said, 'Nonsense!' and went back to sleep—
Jim: That was actually a joke—
Tina: I wasn't questioning your sanity!
Jim: That was a joke. I'd woken up. I just been dreaming and was just a bit out of it still. You keep trying to ask me if I feel like I did when I had the infection—
Tina: Which infection?
Jim: You know when I was behaving strangely. But I just don't know. I don't know if I'm making sense *now*. I assume I am but—
Tina: [tenderly] You're making complete sense—
Jim: But I don't *know—that's the point*. So if I'm having a seizure then I know I'm having a seizure. I don't know how I'm appearing to everyone else. I really don't want to be stuck where everyone else is talking about me and I'm out of it, obviously, knowing what's up but—it's like when I've had the seizures before I can see everyone else's reaction. But I can't talk, which is disturbing. Or when I was about to have one … or when I'm having the aura and I know it's about to happen. And I see other people panic … I can't say, 'stop panicking everyone.' Because I'm shaking and—I think it's weird being aware of what's going on around you when you're having that seizure. Everyone says that you can't—that you should be unaware. But actually each time I've had one I've been awake and aware of what's going on around me. And my arms start moving. And I can see everyone else's reaction. But I can't communicate. It's weird. That's quite unpleasant. That's why I'm so nervous about having one.

A Fate of Confusion

Jim's concern with making sense was by no means unusual. In fact, it was an enduring feature of many accounts. Moreover, it was not simply

12 Living Suspended: Anticipation and Resistance in Brain Cancer

in people's interpersonal relationships, but in their broader relationships with the world around them.

Consider Fay's explanations for a peculiar find in her garden. Fay lived alone after caring for her parents who both died of cancer. Her four sisters would sometimes cram into the consultation rooms at the hospital. After her mother's death two years before, Fay had moved into a smaller place with a small garden. At once her sanctuary and freedom, this garden occupied her days off from the hospital routine and her many family commitments. A year after we met, she told me this story:

> A few weeks ago, I was out here planting some new seeds and I could see this pink thing sticking out of the soil. I thought it was a big fat bud, so I dug around it to see. And Maria [Fay's sister] was with me and I asked her, 'Did you plant a carrot out there? Did you seriously plant a carrot to wind me up?'—because that's the sort of thing Maria would do—but then she said, 'Fay, that's not a carrot; it's a frankfurter!' I'm not kidding! It looked like it had been pile driven straight down into the soil. So, I'm saying to Maria, 'Are you taking the piss?!' I said, 'Maria—,' I said, 'Please don't!'—Because I thought well maybe I planted it and then forgot or— And she said, 'I promise you'—she was rolling around laughing—'Fay, I promise you it wasn't me.' And I said, 'Someone's taking the Mickey,'— you know whether it's the neighbour, or kids, I thought—So she says, 'I wouldn't do that because I know how confusing you find things.' So anyway, a few days go by and I was sitting in the garden and this squirrel came bounding across the shed. And it jumped down into those plants that are in the basket by the wall. Then it lent down, grabbed one of the plants with its little hand and yanked the whole thing up—the whole plant—and then it planted a great big wedge of cheese—I'm not kidding you—a great big wedge of cheese! Then it stood on it and stamped down on it and patted the soil with its hands. And then I'm thinking, okay now I believe it wasn't Maria and it wasn't me: it was the squirrel! And then last week Maria was in the garden and she said, 'Fay, come and look at this.' A whole egg! In its shell! She found a whole egg! And I said how can a squirrel carry an egg? How can he get it over the fence?! I thought I was losing it.

When she told me this, Fay was laughing. She was incredulous at the sight of a squirrel with a frankfurter, a piece of cheese and an egg. But there is a deeper narrative running through the story and this has to do with the alternatives Fay gave by way of explanation: if not Maria or a neighbour's practical joke, perhaps it was her. To anchor the significance of the story, we must return to earlier accounts where she spoke more explicitly about her doubts and how they took shape in her sometimes fraught familial relationships. The following passage is from six months earlier, days after Fay had defended her sovereignty from two of her sisters who, in her words, 'went behind my back,' to arrange a special consultation with the doctors without her:

> *Fay*: It's like I'm not part of my life anymore—I did feel like it wasn't my life. It was like everybody else was making decisions for me. And I said to them, 'I'm not stupid, I'm still here, I'm still Fay, you know!' I felt like they've just taken it out of my hands and it did upset me. I felt betrayed by them both … I just think to myself, 'I've got this, this is happening to me,' and I'm doing my best to—not get over it because you can't get over it—but to get on with it. And I felt when they did that, it kind of put doubt into my mind that I'm not capable of making my own decisions. And last night I lost my keys and I cried my eyes out because I'd thought I'd put them somewhere—my memory isn't great. It was the sheer frustration. I was so upset and it just makes me doubt myself. I think I've put my phone here and I haven't, it's over there. So I'm looking over here for it. And I'm like, 'Where is it? I put it there.' Like I said to Maria this afternoon: people with Alzheimer's, how frightened they must be. Maybe they're clear the one minute and then suddenly they don't know where the hell they are. I don't want that to happen to me. Sometimes I wake up and I think, 'Where the hell am I?' I look around and I think to myself, 'I don't know where I am.' And then it will suddenly come back, 'You've moved [house].' And then I remember everything.
>
> *Henry*: When did you start to have these kinds of thoughts?
>
> *Fay*: Ever since I came out of hospital, but my memory has got worse since I've been home, since I had the radiotherapy and all of that lot, it has been worse. And also because my eyes are so bad. But I don't think, 'Oh maybe I put it somewhere else.' I've convinced myself that's where I put them so I say to myself, 'Why isn't it there?' I don't know what happened last night, I was really upset. Why was I crying over a bunch of keys? … In the future, whether it's a week or you know, five years, I do not want to be dismissed by anyone. Not just the doctors.

> I don't want to be dismissed by my family—not that they would do it deliberately … I just don't want to end up the batty aunt that no one wants to come and see.

Resistance and Continuity

Strategizing ways to counter the frightening anticipation of being lost was also common among participants. These would sometimes be simple affirmations or narrative modes of establishing continuity—as in Fay's, 'I'm still Fay, you know!' and Tina's, 'You're making complete sense, Jim'—at other times, more elaborate and calculated.

George was 66 and recently retired from a life on the railways. His wife, Phoebe, told how this work had a lifelong effect on his sensibilities towards marking time: a 'minute-to-minute' punctuality that kept things precisely in place. Since the tumour, however, George started to lose track of details that ordinarily he took for granted. So, he devised a ledger in tall red book. It read: '0558 up in the morning; headache; 0604 toilet; 0632 made cup of tea; 0708 went back to bed; 1057 woke up again.' This book, which revealed a life measured out in small daily tasks, became a key means of establishing a kind of narrative continuity (Llewellyn et al. 2014) for George and easing his mind. Over time, the log became more sophisticated—different colours, bracketed comments, thoughts, feelings:

> I write down everything and then on the next page I'll write down the things that are relevant to me … it's part of the process [I'm] going through to try and understand it—to have some sort of an idea, a log, to understand how [I've] ended up somewhere.

While waiting to see his oncologist, the three of us sat and at George's request drew a map of the head: 'Then I'll know—I can put 5 for a really bad headache, 1 for less bad.' It was another iteration towards establishing a more accurate account of what George called 'fleeting pains and feelings.' Together, we drew the outline of a head in the red book and divided it into regions where George said he was having headaches. The next time I saw him, he showed me the outline which he had marked with jottings and numbers and described the route his headaches had taken the previous day, moving from front to back and gradually getting less severe. While serving the same base function to establish a memory of events, this was also a technology of communication for George, a bridge between Phoebe, the clinicians and himself: 'That way I can tell them what's what—give them an overall picture they can relate to.'

Others sought different means of establishing continuity and resisting the 'biotechnical truths' they were given in their medical diagnoses. Indeed, the imagery of disease was not always taken as stock and complex interpretations turned biotechnical explanations into new, unexpected and positive framings. This is well illustrated in Jamie's accounts. Jamie was 55 years old when we met, a practicing Buddhist. During daily practice, he would imagine his tumour in elaborate visualizations. For example:

> It's all black—a black blob with white eyes. And it spreads its tentacles … and the tentacles grow and little Buddhas chop them and push them away—chopping them and getting back to the core. And then I've got this group of people who are martial artists. And they've got my tumour in a corner and when it tries to escape they punch and kick it. But then I heard people say that you need to love your tumour. So what the ninjas now do is punch and kick it and keep it in place. And they tie it up. And then they comfort it. And they comfort it and say 'right, you now have to leave this body, sorry, go away.' And then they put it, they literally put it on this set of stairs that goes out to the toilet. And they lock the door and then of course I go to the toilet and flush it away—that's my visualisation—and it goes into the sea.

Another time, Jamie told me that the Buddhas now had motorcycles. He imagined a desert-like scene in the USA and a lost highway stretched out to the horizon. The tumour, bound and gagged, now rode on the back of one of the bikes, flanked either side by a motorcade of Buddhist ninjas and riding off into the distance. His images became more and more elaborate and though he has a particularly brilliant and inventive imagination, they bespeak a serious point about the intertwining of biotechnical imagery, sensation and everyday living, which was common among participants.

Amanda also had strong visualizations. Towards the end of our first interview at her home, I asked her about the scan of a healthy brain she had pinned to a board above her desk, reminiscent of the ultrasound image expectant parents might carry of a child in the womb. She smiled as she said:

> That's not my brain. It's part of my therapy to think positively. I read a book about this—and this really helps. One of the things from this book was to print out the pictures from your dreams—your purpose in the future, what do you want. So I printed a healthy brain. The book said that you should look at this picture every day and visualise this picture and think that this is yours—that this is your dream—and it's going to come true. So when I look at the calendar and see it, I say 'this is going to be my next scan.'

Amanda's and Jamie's accounts are interesting both for their reliance on biotechnical imagery and appropriation of it. In Jamie's case, his visualization transformed the black and white scan into an enemy which he could apprehend and stall through attacks or diplomacy. It gave a story and a logic to his real-world approach. Amanda appropriated the imagery to give substance to her dream and usher forth a more hopeful future. Their visualizations did more than simply make abstract biotechnical truths intelligible: they established a narrative agency, self-actualization, and thereby a kind of mastery in waiting for disease to reassert itself or indeed other futures. Such modes of establishing narrative continuity and resistance thus helped to conjure alternative possibilities and counter the incapacity given in less hopeful visions.

Subjectivity and Immanent Possibility

We have marked above some of the subjective turns produced by changing conceptions of mind: when patients became the subjects of their own doubt and when certain kinds of experience evoked fears of becoming lost. To greater or lesser extents, these turns happened consistently across the patients who Llewellyn met in fieldwork and would recur over the duration of disease. Old memories that imposed themselves, disrupting daily tasks and returning feelings of nostalgia to the present, seem obviously strange and understandably frightening. Yet even things like losing keys that seem innocuous and quotidian on the face of it could induce major anxiety and doubt. While they might be out-of-sync with the habits of everyday life, they are also a normal part of it; usually an explanation soon appears and the uncertainty felt is not typically threatening. But repeatedly experiences—big and small—would be chalked up to the tumour. They would be interpreted as symptoms and narrative proofs. In these narratives, strange events became harbingers of a new fate: 'I really don't want to be stuck where everyone else is talking about me and I'm out of it,' as Jim said.

In his analyses of symptoms, technologies and subjectivities among people cast as mentally ill (Biehl 2010, 2013) and those diagnosed with or at risk from HIV infection (Biehl 2007; Biehl et al. 2001), anthropologist João Biehl deftly weaves the layered productions of people and how they consider themselves. He points to the technoscientific and medical developments in which people and social interactions are constituted and hence the social and technological embedding of subjectivity. Examining HIV public health programmes in Brazil, for example, Biehl highlights the mediating roles of biologically based identities and rational-technical health management concepts, including free HIV tests, which he suggests produce 'a population of ... an *imaginary AIDS*' (Biehl 2007; Biehl et al. 2001, 99, *italics in original*). This population comprises people at low risk of infection who complain of AIDS-like symptoms and demand serial HIV/AIDS testing, only to be returned each time as sero-negative. Biehl suggests that their anxieties and somatic responses are a result of how they have absorbed certain biotechnical truths and developed a morbid and anticipatory subjectivity.

In a particularly insightful passage he discusses what he calls *technoneurosis*—'the confused and painful experiences [that are] somewhat technically engineered' and which establish a kind of 'neurotic "fate"' in people who come to see themselves as being at risk (Biehl 2007, 270). Biehl highlights this as a neurotic disposition that is characterized by the symptoms of an imaginary AIDS rather than a biological reality. He thus draws attention to how certain kinds of predictive profiles are lived as realities by the people they define, even though these people do not have the virus. He therefore sites biotechnology as a 'complex intersubjective actor' among clinicians, policymakers and patients, determinative and capable of producing radical transformations in people's subjectivities (Biehl and Moran-Thomas 2009, 280).

We find this particularly useful in understanding the predictive quality of biotechnical explanations and technologies like brain imaging as they are interpreted by those Llewellyn worked with, all of whom suffered the feelings of what might be termed a 'fate' of confusion. Although we wish not to term these experiences neurotic and while clearly the situations of undiagnosed populations with an imaginary disease and people actually diagnosed with a brain tumour are substantially different, their experiences of inevitability and interpretations of sensations and events as signifying a new imagined reality are strikingly similar. With their fated existences, as they saw them, those Llewellyn met framed new sensations and certain experiences as strange and co-extensive with the biotechnical fact of a tumour, and would each time identify these as harbingers of a new imagined reality—one they could not know. These were experiences that at once embodied and ushered forth a terrifying loss of self.

This brings us to suggest the paradoxical subjectivity particular to people with a brain tumour and perhaps—as Fay indicated in her comparison above—people with neurological diseases, like Alzheimer's. As biotechnical truth is internalized and worked alongside common cultural images of disease it establishes an imaginative space of potential and interpretive frame through which to see experiences and cast them as strange. Patients are led through spiralling doubt into overwhelming frustration and to imagine themselves as the 'batty aunt' or 'out of it'; stuck in a void of their own and dismissed by everyone around them. It is a subjectivity of their own negation: an anticipatory loss of self. The ambiguous nature of this negation is radically disturbing; the very point Jim made when comforted by Tina that he was making sense: '*But I don't know—that's the point.*'

As our analysis shows, however, it is misleading to portray a linear evolution from preclinical explanations through the acquisition of biotechnical truth to an orientation that continually remakes the world conform with diagnosis. Rather, there are multiple frames which mediate the world and render experience continual with a brain tumour, coincidental or discordant. Those Llewellyn met partook in complex and evolving intersubjective spaces and themselves cultivated new imaginative spaces of potential. How they imagined themselves was a constant choreography—new sensations; new explanations; new biotechnical portraits; new relationships and ways of relating; narrative oscillations; new affirmations; new ways to normalize; new appropriations; new improvised techniques and more sophisticated means of mending shared realities. It is this choreography which we site as a form of active waiting; not hopeful orientations or modes of establishing an imagined future more quickly, but the tactical modes to postpone the terrifying images of mental incapacity.

Conclusion

The lives of those suffering progressive disease are marked by radical turns and discontinuities. And yet they are also marked by suspension. In this chapter, we have documented these moments of suspension and the imagined possibilities that proliferate therein. In particular, we detailed the development of what we call an anticipatory loss of self: a subjectivity and an interpretation of oneself—mind and brain—produced in the intersubjective encounters between people with a brain tumour, their families, clinicians, biomedical technologies and the physical sensations that emerge as tumours develop and patients undergo intensive monitoring, surgery and therapies. This anticipatory loss of self is not merely an emotional state, but a mode of being that establishes a frightening imagined reality and through which one questions physical sensations, the nature of shared reality and ultimately one's own capacity as a rational agent. We have also documented tactical modes by which patients attempted to forestall terrifying and profoundly unwanted futures. These attempts could be literal, via treatment, relational, through attempts to mend a shared sense of reality, and figurative, in the appropriation of biotechnical imagery. Analytically, we consider these modes as active forms of waiting different to previously documented ideas of hope and intention, and the means of actualizing potential more quickly. We argue that such focus on living in suspension and the imagined possibilities of becoming helps us better understand the dynamics of stasis and flux during progressive disease and ultimately what it is to 'live in prognosis' (Jain 2013).

* * *

Llewellyn interviewed and spent time with sixteen people with a brain tumour and often members of their families. Of those people quoted here, Fay was admitted to hospital after a catastrophic seizure and then entered a hospice where she died. Sara died soon after finishing chemotherapy treatment, which she stopped early due to overwhelming side effects. George died at home and during fieldwork. Jim, Jamie and Amanda survive at the time of writing.

References

Andersson, Ruben. 2014. "Time and the Migrant Other: European Border Controls and the Temporal Economics of Illegality." *American Anthropologist* 116 (4): 795–809. https://doi.org/10.1111/aman.12148.

Appadurai, Arjun. 2013. *The Future as Cultural Fact: Essays on the Global Condition*. London: Verso.

Auyero, Javier. 2012. *Patients of the State: The Politics of Waiting in Argentina*. London: Duke University Press. https://doi.org/10.1111/amet.12111.

Biehl, João. 2007. *Will to Live: AIDS Therapies and the Politics of Survival*. Princeton, NJ: Princeton University Press.

———. 2010. "Human Pharmakon: Symptoms, Technologies, Subjectivities." In *A Reader in Medical Anthropology: Theoretical Trajectories, Emergent Realities*, edited by Byron J. Good, 213–231. Oxford: Blackwell.

———. 2013. *Vita: Life in a Zone of Social Abandonment*. London: University of California Press.

Biehl, João, and Amy Moran-Thomas. 2009. "Symptom: Subjectivities, Social Ills, Technologies." *Annual Review of Anthropology* 38: 267–288. https://doi.org/10.1146/annurev-anthro-091908-164420.

Biehl, João, Denise Coutinho, and Ana Luzia Outeiro. 2001. "Technology and Affect: HIV/AIDS Testing in Brazil." *Culture, Medicine and Psychiatry* 25 (1): 87–129.

Biehl, João, and Peter Locke. 2010. "Deleuze and the Anthropology of Becoming." *Current Anthropology* 51 (3): 317–351. https://doi.org/10.1086/651466.

———. 2017. "Introduction: Ethnographic Sensorium." In *Unfinished: The Anthropology of Becoming*, edited by João Biehl and Peter Locke, 1–40. London: Duke University Press.

DelVecchio Good, Mary-Jo, Byron J. Good, Cynthia Schaffer, and Stuart E. Lind. 1990. "American Oncology and the Discourse on Hope." *Culture, Medicine and Psychiatry* 14: 59–79.

DelVecchio Good, Mary-Jo, Tseunetsugu Munakata, Yasuki Kobayashi, Cheryl Mattingly, Byron J. Good, and Peter Brooks. 1994. "Oncology and Narrative Time." *Social Science & Medicine* 38 (6): 855–862.

Dwyer, Peter D. 2009. "Worlds of Waiting." In *Waiting*, edited by Ghassan Hage, 15–26. Melbourne: Melbourne University Press.

Ehn, Billy, and Orvar Löfgren. 2010. *The Secret World of Doing Nothing*. Berkeley: University of California Press.

Fox, Renée C. 2000. "Medical Uncertainty Revisited." In *The Handbook of Social Studies in Health and Medicine*, edited by Gary L. Albrecht, Ray Fitzpatrick, and Susan C. Scrimshaw, 413–439. London: Sage. https://doi.org/10.4324/9780203996751-22.

Hage, Ghassan, ed. 2009. *Waiting*. Melbourne: Melbourne University Press.

Jain, S. Lochlann. 2013. *Malignant: How Cancer Becomes Us*. London: University of California Press.

Janeja, Manpreet K., and Andreas Bandak, eds. 2018. *Ethnographies of Waiting: Doubt, Hope and Uncertainty*. London: Bloomsbury.

Llewellyn, Henry, Joe Low, Glenn Smith, Katherine Hopkins, Aine Burns, and Louise Jones. 2014. "Narratives of Continuity Among Older People with Late Stage Chronic Kidney Disease Who Decline Dialysis." *Social Science & Medicine* 114: 49–56.

Llewellyn, Henry, Paul Higgs, Elizabeth L. Sampson, Louise Jones, and Lewis Thorne. 2018. "Topographies of 'Care Pathways' and 'Healthscapes:' Reconsidering the Multiple Journeys of People with a Brain Tumour." *Sociology of Health & Illness* 40 (3): 410–425. https://doi.org/10.1111/1467-9566.12630.

Louis, David N., Arie Perry, Guido Reifenberger, Andreas von Deimling, Dominique Figarella-Branger, Webster K. Cavenee, Hiroko Ohgaki, Otmar

D. Wiestler, Paul Kleihues, and David W. Ellison. 2016. "The 2016 World Health Organization Classification of Tumors of the Central Nervous System: A Summary." *Acta Neuropathologica* 131 (6): 803–820. https://doi.org/10.1007/s00401-016-1545-1.

Mattingly, Cheryl. 2019. "Waiting: Anticipation and Episodic Time." *The Cambridge Journal of Anthropology* 37 (1): 17–31. https://doi.org/10.3167/cja.2019.370103.

Petryna, Adriana. 2002. *Life Exposed: Biological Citizens After Chernobyl*. Princeton, NJ: Princeton University Press.

Reed, Adam. 2011. "Hope on Remand." *Journal of the Royal Anthropological Institute* 17: 527–544.

Rose, Nikolas, and Joelle M. Abi-Rached. 2013. *Neuro: The New Brain Sciences and the Management of the Mind*. Princeton, NJ: Princeton University Press.

Stephan, Christopher, and Devin Flaherty. 2019. "Introduction: Experiencing Anticipation: Anthropological Perspectives." *The Cambridge Journal of Anthropology* 37 (1): 1–16. https://doi.org/10.3167/cja.2019.370102.

Taussig, Karen-sue, Klaus Hoeyer, and Stefan Helmreich. 2013. "The Anthropology of Potentiality in Biomedicine An Introduction to Supplement 7." *Current Anthropology* 54 (S7): S3–14. https://doi.org/10.1086/671401.

The Brain Tumour Charity. 2015. "Losing Myself: The Reality of Life with a Brain Tumour."

Vidal, Fernando. 2009. "Brainhood, Anthropological Figure of Modernity." *History of the Human Sciences* 22 (1): 5–36. https://doi.org/10.1177/0952695108099133.

Williams, Simon J., Paul Higgs, and Stephen Katz. 2012. "Neuroculture, Active Ageing and the 'Older Brain': Problems, Promises and Prospects." *Sociology of Health & Illness* 34 (1): 64–78. https://doi.org/10.1111/j.1467-9566.2011.01364.x.

Index

A

acute medical unit 208, 214, 216, 218
admission 190, 207–209, 212, 213, 223
affective 4, 10, 56, 131, 198, 252, 253, 258
Afghanistan 41
algorithm 6, 16, 17, 22, 24–27, 31, 33
Alzheimer's 208, 209, 222, 267
ancestor 113–119, 121–127, 130–132
anticipation 8, 48, 53, 71, 76, 252, 263
anxiety 5, 55, 67, 79, 137, 159, 212, 244, 266
Argentina 5, 45, 62–66, 68, 74, 85–87, 93, 101, 103
asylum 40–43, 45–50, 53–55

Athens 42

B

body 6, 7, 10, 23, 25, 26, 55, 64–68, 73, 74, 78–80, 88, 102, 114, 117, 119–121, 124, 130, 136–142, 144–150, 156–164, 166–175, 177, 211, 217–219, 221, 222, 251
border 40, 41, 234
brain injury 155
Buenos Aires 61, 63, 64, 66–68, 71, 75, 86, 87, 96, 99, 102
bureaucracy 18, 97, 98

C

cancer 5, 6, 18, 23, 27–29, 85–87, 90, 92–95, 101–103, 232,

Index

236, 237, 241, 242, 244, 251, 255, 261
care 2, 4, 5, 7–9, 63, 69, 74–77, 85, 86, 89, 90, 93, 95, 97, 101, 102, 137, 139–142, 144–147, 149, 150, 164, 186–188, 191–193, 195, 202, 207–209, 211, 212, 214, 218, 223, 232, 234–236, 240, 241, 243–245, 251, 252, 254–256
caregiving 187
childhood 87, 90, 93, 101, 217, 236, 258
children 8, 30, 46, 47, 49–51, 55, 78, 85–87, 90, 95, 99, 101, 102, 123, 126, 191, 193, 195, 202, 232, 233, 236–245
chronic disease 87, 89, 94, 95, 100
classification 25
clinician 8, 63, 146, 241, 252, 254, 255, 264, 267, 268
collaborative 42, 238
confusion 159, 258, 267
critical phenomenology 187, 189, 197, 202
culture 118, 120, 129, 132, 209, 211, 255, 257

D

dementia 5, 7, 186–194, 198, 201, 203, 207–215, 218, 221–223
dependency 17, 30, 34, 136, 140, 189, 209
depression 4, 46, 137, 171
disability 137, 141, 143, 146, 147, 149, 157, 161–163, 169, 172, 173, 189, 209, 210

discourse 7, 20, 42, 75, 79, 80, 149, 161, 164, 191
disease 6–8, 19, 22, 27, 28, 86–91, 93–99, 101–103, 155, 163, 177, 190, 203, 242, 243, 252, 254–256, 264–269
donor 15, 16, 18–26, 29–33

E

embodiment 6, 66
empowerment 4
endocrinologist 63, 77
environment 7, 65, 89, 98, 130, 141, 143–145, 156, 157, 159, 160, 162–164, 171, 172, 177, 191, 194, 210–212, 218
epoch 141
ethnography 100, 118
European Union (EU) 40–42, 49
Eurotransplant 18, 22, 24–26
existence 44, 86, 121, 125, 138, 141, 162, 188–190, 195, 197, 203, 267
expectant waiting 20, 21, 32, 34

F

farmworkers 8, 231–238, 240–245
fieldwork 87, 113, 118, 119, 122, 253, 254, 257, 266, 269
Florida 231, 237, 238, 241
fragility 32, 34

G

gender identity 62, 64, 65, 73, 75, 80
Gender Identity Law 62, 64, 68, 71, 74, 78, 80

generic waiting 67
German Association for Organ Donation (DSO) 24
Germany 5, 16–19, 21, 26, 52, 53
Greece 5, 40–44, 46–56

H

handicap 143
healing 101, 102, 116, 123, 127, 132, 146
healthcare 2, 7, 9, 10, 18, 20, 21, 28, 62, 63, 65, 66, 68, 71, 72, 74, 77–80, 102, 145, 147–150, 232, 234, 235, 238, 239, 242, 245
homeland 115–118, 123, 126–128, 130, 131
hope-generating machine 16, 18, 20, 31, 33, 34
hospital 16, 18, 63, 65, 66, 68, 71, 74–76, 79, 86, 87, 93, 94, 98, 99, 139, 140, 162, 207–214, 221–223, 254, 257, 258, 261, 269
hospitalization 208
household 23, 113, 114, 117, 120, 122, 123, 125–127, 187
Human Papilloma Virus (HPV) 8, 232, 233, 235–238, 240–245

I

identity 7, 31, 43, 63, 65, 74, 75, 114, 121, 123, 129, 156, 158–163, 169–173, 175, 176, 178, 192, 266
illness 2–4, 10, 85, 88, 90, 92, 93, 96–98, 102, 103, 118–120, 126, 130, 131, 155, 156, 158–160, 162, 177, 197
imagined possibilities 8, 252, 268, 269
immobility 2–10, 16, 17, 20, 30–32, 34, 44, 46, 55, 56, 86, 87, 91, 93, 95, 98–103, 115, 116, 118, 121, 126, 131, 132, 136–139, 141, 148, 155, 156, 159–162, 164, 165, 167–171, 176, 186, 189, 202, 203, 208–212, 222, 233
in-between 10, 76, 91, 102, 149, 188
in-betweenness 3, 4, 7, 149
in-depth interviews (IDI) 42, 43, 63, 87
individual movement 4, 6, 41, 161
inequality 9, 74, 88, 91, 102, 129, 235, 244
infrastructure 5, 9, 16, 18, 21, 23, 27, 33, 34, 71, 101
interviews 17, 42–44, 48, 63, 68, 71, 73–75, 77–79, 124, 166, 213, 215, 238, 254, 265

L

lifespan 140, 141
liminality 3, 6, 9, 19, 40, 87, 91–93, 101–103, 136, 138, 143, 144
list 6, 16–18, 21–25, 27, 28, 33
loss of self 8, 254, 257, 267, 268

M

material culture 3, 6
materiality 3, 16, 23, 24

medical care 17, 20, 28, 73, 76, 77, 86, 87, 103, 208
medical institution 63
medicalization 75, 190
medicine 2–4, 7, 16–21, 24–28, 31, 33, 34, 75, 94, 97, 102, 114, 117, 119, 121, 137, 138, 142, 143, 147, 252, 254, 255
memory 74, 187, 190, 213, 215, 217, 258, 264, 266
mental disorder 64
mental health 42, 43, 53, 54, 137
mental health distress 40
mental health services 43, 234
migrant health 238
migration 20, 21, 40, 86, 87, 90, 91, 93, 95, 97, 98, 101, 103, 117, 118, 122, 123, 127–130, 240
migratory labour 235, 244, 245
mobile phones 15–17, 23, 30–34, 115
mobility 2–4, 6, 7, 9, 10, 16, 17, 31, 32, 34, 41, 49, 55, 56, 80, 86, 90, 91, 99, 101, 102, 115, 138, 140, 141, 143, 156, 158, 159, 161, 163, 164, 167, 169, 170, 172–178, 186, 187, 189, 190, 201, 203, 208–212, 214, 218, 220, 222, 233, 243, 244, 246
Model of End-Stage Liver Disease (MELD) 26–29, 33
moment 2, 8, 9, 22, 30, 47, 51, 68, 70, 91, 92, 94, 100, 102, 156, 158, 159, 162, 166, 185–189, 192, 196, 199–202, 219, 232, 252, 253, 255, 258, 268
motility 3, 4, 8, 9, 100, 156–158, 164, 233–235, 243–245

N
neurology 193
Neuro-oncology 251
non-governmental organization (NGO) 43, 87

O
organ 18–26, 29, 31, 33, 136, 137, 140

P
paralysis 120, 121, 136, 142, 166
participant observation 87, 254
passivity 41, 44, 45, 55, 141
pathology 64, 147, 255
patient 2, 4–8, 16–29, 31–34, 62, 63, 66–71, 74–79, 85, 86, 89, 90, 93, 102, 136, 141–146, 149, 150, 159, 163, 164, 177, 207–213, 216, 221–223, 236, 238, 241, 251–258, 266–268
perception 7, 79, 90, 98, 99, 163–166, 169, 173–175, 177, 222
period 21, 30, 61, 66, 67, 69, 71–73, 75, 76, 78, 79, 86, 87, 90, 93, 95, 97, 98, 100, 102, 103, 116, 118, 119, 122, 125, 127, 140, 141, 187, 231, 251, 255
Phenomenology 157
physician 21, 24, 28, 29, 136, 139, 142–146, 149
physiology 174
policy 49, 61, 62, 88, 102, 115, 123, 191, 232, 234, 235, 237, 238, 244, 245, 254

political economy 118, 123, 124, 128, 129, 235
potential 4, 8, 19, 20, 25–27, 32, 34, 47, 48, 90, 92, 93, 155–157, 161, 174, 189, 193, 197, 198, 233, 235, 236, 243, 252, 267–269
potentiality 187, 197, 252, 254, 257
power 4, 5, 8, 19, 48, 54, 62, 66, 67, 72, 78, 79, 88, 115, 120, 123, 128, 146, 147, 208, 213, 259
power relations 6, 9, 20, 66, 80, 91
praxis 115, 116
prevention 88, 139, 189, 241, 243
psychologist 32, 63, 74
psychosocial interventions 54, 56
Public Health Service 68

R
refugee 5, 40–51, 53–56, 253
refugee camps 5, 9, 41–44, 53–55
resistance 7, 129, 149, 209, 265
responsive engagement 187, 201, 203
rites of passage 91–93

S
Samos 42, 43, 48, 49
scenes 64, 66, 71, 73, 74
SCI body 7, 138, 140–142, 144, 149
self 7, 65, 115, 148, 156, 158, 160–163, 169, 173, 175, 176, 221, 252, 256, 257, 265
sickness 119
socialization 64

social navigation 156
South Africa 5, 113, 115–118, 124–126, 128–130, 132
Spinal Cord Injury (SCI) 5, 7, 136–139, 141, 142, 144, 145, 147–149
spirit possession 115–119, 127, 130, 131
stillness 3, 4, 7–9, 141, 187, 188, 194–196, 198, 199, 201, 202, 222, 243
stroke 5, 7, 155, 156, 161, 163–167, 170–177, 259
structural barriers 8, 233, 237, 238
structure 3, 5, 44, 53, 62, 78, 79, 88, 89, 91–93, 97, 101, 102, 119, 126, 143, 145, 150, 156
stuck 2, 4, 6, 19, 30, 40, 47, 49, 132, 136, 139, 140, 142, 144–146, 148, 149, 159, 165–167, 211, 222, 233, 234, 243, 266, 267
stuckedness 4, 7, 10, 20, 30, 32, 34, 40, 46, 49, 115, 136, 137, 212, 222
subjectivity 8, 21, 88, 252, 253, 266–268
suspension 252, 268, 269

T
technology 2, 6, 8, 16, 17, 21, 23, 26–28, 30, 34, 102, 114, 252, 264, 266–268
technoneurosis 266
temporality 16, 62, 66, 77–80, 233
therapy 8, 66, 209, 268
transgender 5, 62–69, 71–75, 77–80
transplant 5, 6, 15–34, 99

trauma 53, 136, 138, 139, 148, 197
treatment 5, 6, 9, 16–24, 28, 29, 31, 33, 63, 65, 73, 74, 76, 78, 79, 85–87, 93, 95, 98–101, 177, 211–213, 251, 252, 254, 255, 268, 269
Tshikuani 114–119, 121, 123–127, 130, 131
tumour 8, 26, 27, 29, 96, 251, 252, 254, 256, 257, 263, 264, 266–269
Turner, Victor 9, 91, 92, 100, 123

U

uncertainty 2, 3, 6, 19, 20, 43, 46–49, 55, 87, 89, 95, 97, 100, 102, 103, 170, 174, 254, 266

V

vaccination 8, 232, 233, 235–238, 241–244
Van Gennep, Arnold 91, 92
Venda 113–119, 122–127, 130–132
Visual methodologies 214

W

waiting 2–6, 8, 10, 16–21, 23, 24, 26, 28, 30–34, 40, 41, 43–45, 47–56, 61–63, 65–68, 71–73, 75, 76, 78, 79, 99, 187, 232, 234, 252, 253, 255, 263, 265, 268, 269
waiting experience 5, 16, 21, 33, 62–64, 66, 67, 71, 73, 76, 78–80, 253
waiting lists 6, 9, 16, 17, 19, 21–25, 27, 29–31, 33, 34

CPI Antony Rowe
Chippenham, UK
2020-10-05 20:13